This book is dedicated to the many good, sincere students I have taught in my biology and life science classes over the years. Much of the material in this book has been tried out on them!

CONTENTS

CONTENTS

CONTENTS

CONTENTS

PREFACE

This book is for people who want to get acquainted with the concepts of basic biology without taking a formal course. It can serve as a supplemental text in a classroom, tutored, or home-schooling environment. It should also be useful for career changers who need to refresh their knowledge of the subject. I recommend that you start at the beginning of this book and go straight through.

This book seeks to give you an intuitive grasp of biology and its terminology. It helps do this by presenting you with 4 icons (cartoon pictures) representing various facets of Biological Order, versus Biological Disorder. Key facts within these categories are numbered in the page margins. You are encouraged to briefly rewrite these key facts in the blanks provided at the end of each chapter.

This introductory work also contains an abundance of practice quiz, test, and exam questions. They are all multiple-choice, and are similar to the sorts of questions used in standardized tests. There is a short quiz at the end of every chapter. The quizzes are "open-book." You may (and should) refer to the chapter texts when taking them. When you think you're ready, take the quiz, write down your answers, and then give your list of answers to a friend. Have the friend tell you your score, but not which questions you got wrong. The answers are listed in the back of the book. Stick with a chapter until you get most of the answers correct.

This book is divided into 4 sections. At the end of each section is a multiple-choice test. Take these tests when you're done with the respective sections and have taken all the chapter quizzes. The section tests are "closed-book," but the questions are not as difficult as those in the quizzes. A satisfactory score is three-quarters of the answers correct. Again, answers are in the back of the book.

There is a final exam at the end of this course. It contains questions drawn uniformly from all the chapters in the book. Take it when you have finished all four sections, all four section tests, and all of the chapter quizzes. A satisfactory score is at least 75 percent correct answers.

With the section tests and the final exam, as with the quizzes, have a friend tell you your score without letting you know which questions you missed. That way, you will not subconsciously memorize the answers. You can check to see where your knowledge is strong and where it is not.

I recommend that you complete one chapter a week. An hour or two daily ought to be enough time for this. When you're done with the course, you can use this book, with its comprehensive index, as a permanent reference.

Suggestions for future editions are welcome.

Now, work hard! But, be sure to have fun! Best wishes for your success.

The Hon. Dr. Dale Pierre Layman, PhD

Acknowledgments

Illustrations in this book were generated with *CorelDRAW*. Some clip art is courtesy of Corel Corporation, 1600 Carling Avenue, Ottawa, Ontario, Canada K1Z 8R7.

I extend thanks to Emma Previato of Boston University, who helped with the technical editing of the manuscript for this book.

I also wish to specially thank Mr. Scott Grillo (publisher) for his continuing interest, as well as Stan Gibilisco, the Series Editor.

Angela Lopez, a former student of mine, has worked hard with me to create illustrations that clearly embody the concepts of biology. We hope you enjoy them!

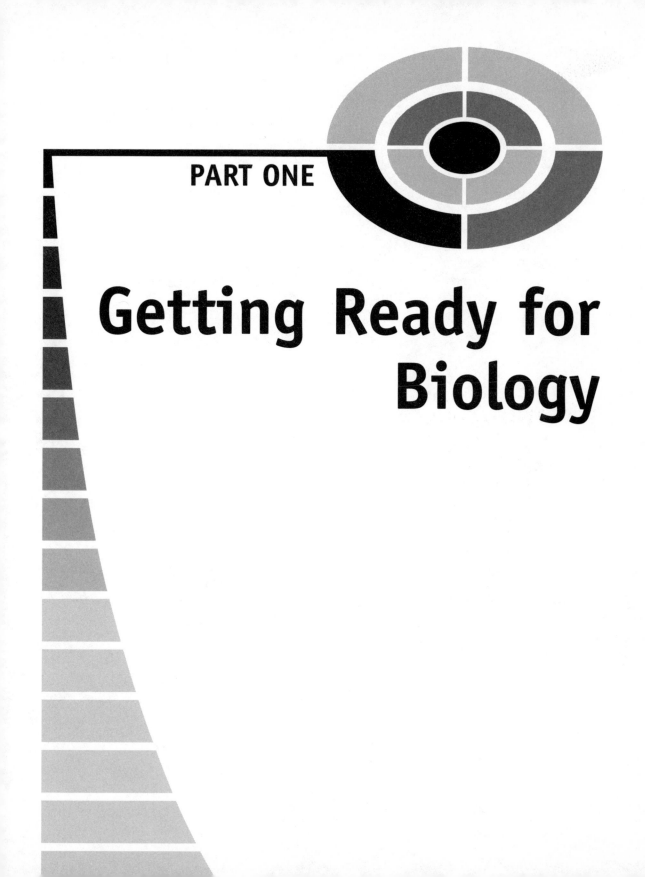

PART ONE

Getting Ready for Biology

1

The Coming of Biology

Hello, there! It's time for you to start teaching yourself some of the fundamental and fascinating facts about *biology* (pronounced as buy-**AHL**-oh-jee). The word, biology, comes from the ancient Greek *bi* ("life") and *-ology* ("study of"). Modern biology, then, is literally the "study of life" in all its forms – plant, animal, or otherwise.

Biology, Organisms, and Order

Biology as the scientific study of living things is a relatively recent development. The term has been in use since the early 1800s to describe the study of living *organisms* (pronounced as **OR**-gan-izms). Basically, an organism is a living thing that is highly organized. By this we mean that an organism has a high degree of *Biological Order*, a very organized *pattern*. (Observe the patterns of the organisms shown in Figure 1.1.)

1, Order

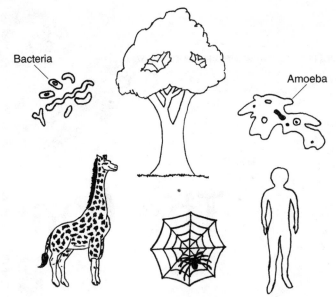

Fig. 1.1   Organisms: Large and small living things with order

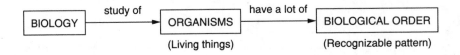

Structure and Anatomy

2, Order

All organisms have some type of *structure*. A structure is literally a "building" up of something from a number of different parts. The structure of an organism, therefore, is the arrangement or pattern "built" up from its individual body parts. The head, neck, and black spots of a giraffe, for instance, help "build" up its entire body pattern.

Anatomy (**ah-NAT**-oh-me) is body structure and the study of body structures. Every kind of organism has its own unique type of anatomy (body structure). When this anatomy is *internal*, being located "inside," it is studied by dissecting or cutting up the body. This reveals why the word, anatomy, exactly means "the process of cutting" (*-tomy*) the body "up or apart" (*ana-*). The internal anatomy of a frog, for example, is revealed whenever its body is dissected or cut apart.

Gross versus Microscopic Anatomy

There are two basic subdivisions of anatomy. These are *gross anatomy*, which studies large body structures (visible to the naked eye), and *microscopic* (**my-kroh-SKAHP**-ik) *anatomy*, which studies "tiny" (*micro-*) body structures (invisible to the naked eye). [**Study suggestion:** Carefully review Figure 1.1. Which of the organisms pictured represent cases of gross anatomy? Which represent microscopic anatomy?]

Both *bacteria* (bak-**TEE**-ree-ah) and the *amoeba* (ah-**ME**-bah) can be classified as *microbes* (**MY**-krohbs). A microbe is a "tiny" (*micr*) "living" (*-obe*) thing. Of course, microbes and other examples of microscopic anatomy can only be examined through a microscope.

A dead giraffe still has gross anatomy. And a dead microbe still has microscopic anatomy. Hence, anatomy occurs in both living and dead organisms (provided they have not decomposed).

Function and Physiology

Just as all organisms have a certain type of body structure (anatomy), they also carry out various *functions*. A function is some type of "performance." It is something that a structure does, or something that is done to a particular structure. For example, both a hammer and a nail are structures. When a hammer pounds a nail (or when a nail is being pounded by a hammer), a function or "performance" is being carried out. The hammer is doing something (pounding the nail), and the nail is having something done to it (being pounded upon).

You might consider function as being a verb, because it involves some action. Conversely, you could think of structure as being a noun, because it is some thing.

When we consider living organisms, the word, *physiology* (**fih**-zee-**AHL**-uh-jee), is used. Physiology is body function and the study of body functions. Unlike anatomy, physiology only occurs within living organisms. Why? Only living organisms carry out body functions. A dead, pickled frog, for instance, still has a recognizable anatomy. It has both legs still present, and a heart. But its legs are limp and lifeless, and the heart pumps no blood. Thus, the frog's body no longer has physiology, because it is dead.

Characteristics of Living Things

Since physiology (body function) only occurs within living organisms, it is very important for us to know the basic characteristics of living things:

3, Order

1. Living things have a high degree of Biological Order (body pattern or organization). In general, living organisms are much more organized than dead ones. And we have already said that living things have an extremely high degree of Biological Order. In the human body, for instance, our *oral* (**OR**-al) body temperature taken by "mouth" (*or*) remains relatively constant at about 98.6 degrees Fahrenheit. Body temperature goes up and down, but it still remains within a *normal range*. It only goes up about 1 degree Fahrenheit, or down about 1 degree Fahrenheit, from the 98.6 degree level. This creates a roughly S-shaped pattern over time (Figure 1.2, A).

Similar S-shaped patterns could be constructed for such aspects of physiology as heart rate and respiratory rate (rate of breathing), and such aspects of anatomy as blood sugar concentration and bone density. In general,

In this book, we will use the very familiar pattern of black spots on a living giraffe (Figure 1.2, B) as a symbol to represent particular cases of Biological Order within living things.

2. Living organisms are sensitive to changes in their surrounding environment and respond to them. An earthworm will respond to hot, dry air by quickly burrowing into the cool, moist earth for safety. A row of sprouting bean plants in a box by a window will eventually tilt and grow in the direction of the incoming light. These are just two of the ways in which the huge variety of living organisms can respond to changes in their surrounding environment.

3. Living things produce movement, either internally or externally (outside of their bodies). A crab, for instance, moves its claws *externally* to clasp a dead minnow and engulf it. The tiny pieces of eaten minnow then move internally, through the crab's digestive tract.

4. Living organisms undergo growth and specialization of their anatomy and physiology as they become older. At first, a newborn shark is tiny, and it has few or no teeth in its jaws. But the shark keeps growing larger and larger for as long as it lives. And it develops a set of razor-sharp teeth in both jaws for cutting and shredding its prey.

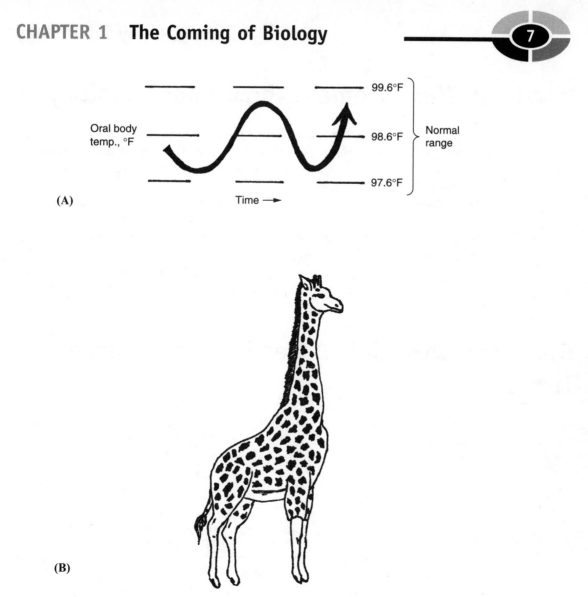

(A)

Oral body temp., °F

99.6°F

98.6°F

97.6°F

Normal range

Time →

(B)

Fig. 1.2   Some patterns representing biological order.

5. Living organisms undergo metabolism for energy production and excrete unuseable waste products. The word *metabolism* (meh-**TAB**-ah-**lizm**) means "a state of change." In living things, food that is eaten is soon changed by the chemical processes of metabolism. Energy is produced, which then performs body work. *Excretion* (**eks-KREE**-shun) is a "sifting out" of waste matter after metabolism, thereby eliminating it from the body.

6. Living organisms reproduce themselves. A living creature has the ability to reproduce itself, either *sexually* (using sex) or *asexually* (**ay-SEK**-shoe-ah-lee) – "not" (*a-*) using sex.

What Did The Cavemen Know About Biology?

Apparently, the very earliest people knew a lot about biology! Ancient caves have been discovered in various parts of the world. The walls of many of these caves are richly decorated with pictures of plants, birds, animals, and the humans hunting them.

Over 2000 years ago, the Early Egyptians brushed colorful paintings of living things onto the walls of their temples and coffins. It must have been very important for these people to begin The Search for Biological Order, a search for predictability and meaning in their natural environment. By discovering the underlying patterns of plant and animal structure, function, and behavior, early humans could ensure a more consistent supply of food.

Biological Disorder: A Reality Since Ancient Times

Earlier, we said that a state of Biological Order (particular patterns associated with living things) was an important property of organisms. It is only logical, therefore, for us to assume that ancient people were likewise aware of its exact opposite – *Biological Disorder*:

1, Disorder

| BIOLOGICAL DISORDER | — | BREAKING A PARTICULAR PATTERN ASSOCIATED WITH LIVING THINGS |

4, Order

2, Disorder

Recall that Figure 1.2 showed some patterns representing Biological Order: There was a giraffe with its intricate pattern of black spots, and an S-shaped curve represented a normal range where something (like oral body temperature) remained relatively constant over time. In a dramatically opposite way, Figure 1.3 (A) represents Biological Disorder symbolically as a dead giraffe without its spots! And Figure 1.3 (B) reveals that a disordered state of something like oral body temperature would be represented by a curve that rose either significantly above its normal range, or significantly below it.

In general, a condition of Biological Order is associated with health and lack of injury. Conversely, a condition of Biological Disorder is often associated with disease, injury, or death of the organism.

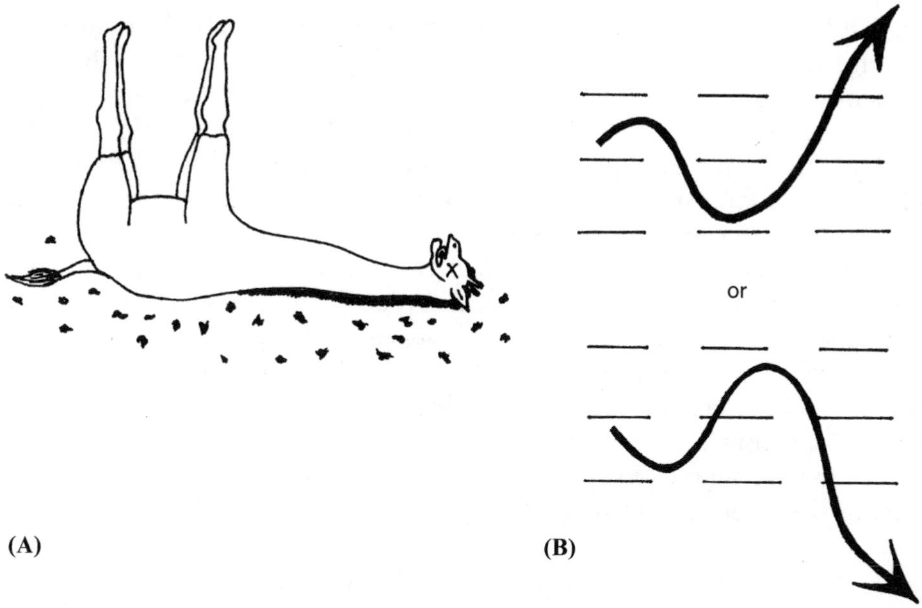

(A)   (B)

Fig. 1.3   Some broken patterns representing biological disorder.

A healthy long bone, for example, has a high degree of Biological Order. But a fractured long bone, in marked contrast, reflects a lot of Biological Disorder. From the very earliest days of human existence, bone fractures have been very common. Hence, surely mankind has always been aware of Biological Disorder (broken patterns) of various kinds, even though ancient people may not always have been consciously thinking about it.

(**NOTE:** As you have already seen, key facts describing some aspect of Biological Order for an individual organism will usually be tagged by the spotted giraffe icon. And key facts describing a feature of Biological Disorder will be tagged by a dead giraffe that has lost its spots.)

Order beyond the Individual Organism: The Spider and its Web

Beyond the internal environment within a particular organism, lies the *external* (**eks-TER**-nal) environment – the surrounding area "outside of" the body. The external environment basically involves patterns in *ecology* (e-**KAHL**-oh-**jee**), including various *ecological* (e-koh-**LAHJ**-ih-kal) *relation-*

ships. Ecology exactly translates to mean "study of household affairs" in common English. Ecology is the study of the relationships between different organisms, and the "affairs" between these organisms and the total "household" of the external environment. Ecology would not focus upon just an individual giraffe, for example. Rather, it would expand its focus beyond the single giraffe organism, to other nearby giraffes, other species, and even the entire external environment of the African plain.

We might picture ecology (ecological relationships) as a vast, interconnected spider web. Therefore, Biological Order in the external environment beyond the organism reflects a healthy or successful pattern of ecological relationships. We will represent this Order symbolically by a spider in its web (Figure 1.4, A). Conversely, Biological Disorder beyond the organism reflects a breaking of the normal pattern of ecological relationships. This state of environmental Disorder is indicated by the icon of a dead spider and its broken web (Figure 1.4, B).

1, Web

1, B-Web

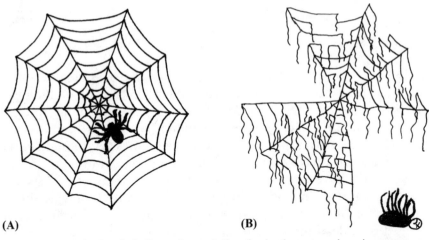

(A)   (B)

Fig. 1.4 Symbols for order and disorder in the external environment.

Natural History: The Search for Order Begins in Earnest

Although ancient peoples around the world have been informally learning about Biological Order (and its counterpart, Biological Disorder) for thousands of years, a systematic approach to the search for Order has only begun relatively recently. *Aristotle* (**AIR**-ist-ahtl), who lived during 384–322 B.C. in

Greece, was one of the first *natural philosophers* (Figure 1.5). Aristotle may have been the world's first great biologist. He collected huge amounts of information about the anatomy, physiology, and behavior of many plants and animals. He was the first person to describe the differences between plants and animals, and between living versus non-living things. Aristotle studied the parts of living organisms in terms of their *teleology* (**tel**-e-**AHL**-oh-jee), their "complete" (*teleo-*) or final purpose for existence. The final "purpose" of numerous plants and animals, this early biologist thought, was to provide food for mankind, thereby adding to human happiness. Aristotle emphasized the importance of making detailed observations to help state general principles, and he assumed that the World of Nature had a great deal of Order within it. Aristotle may also be considered the Father of Natural History – the collection and classification of plants and animals into particular groups.

Aristotle's ideas about a broad Natural History – the comprehensive study of plants, animals, fish, birds, and all other living creatures – were widely adopted and followed for hundreds of years. During the 1700s in Europe, a number of brave explorers made dangerous journeys to faraway places around the globe, collecting and classifying a great variety of plants and animals. Gradually, the immense Biological Order of the world was becoming quite apparent. The ability of naturalists to keep classifying newly discovered organisms into particular groups was strong evidence of a high degree of Order underlying the natural environment.

2, Web

Fig. 1.5   Aristotle: The world's first great biologist.

The 1800s: Experimental Physiology Comes of Age

By the 1800s, the collecting and classifying activities of Natural History were becoming more and more replaced by the *experimental method*. In this method, an *hypothesis* (high-**PAHTH**-eh-sis) or starting hunch is "placed" (*thesis*) "below" (*hypo-*) a group of facts as a tentative explanation. The researcher then conducts an *experiment* to test the hypothesis or hunch and collects some results. The results are compared to what was originally predicted by the hypothesis. A *conclusion* is reached. Either the hypothesis is accepted (being supported by the collected evidence) or it is rejected as not being consistent with the evidence. By this more sophisticated experimental method, no particular hypothesis is ever proven as "absolutely true." There is always the possibility that a new experiment may be conducted which will provide contradictory evidence. Nevertheless, the experimental method began to provide ever-closer and closer fitting to the "real truth" as it exists in our natural environment.

One of the key investigators using the experimental method in the mid-1800s was a Frenchman named *Claude Bernard* (ber-**NAR**). Claude Bernard was one of the first *experimental physiologists* (fiz-e-**AH**-loh-**jists**). As "one who studies" (*-ologist*) about body "Nature or function" (*physi*), Bernard performed lots of experiments on dogs. Bernard did much of the early work on *homeostasis* (**hoh**-mee-oh-**STAY**-sis). Homeostasis literally means a "control of sameness," that is, a relative constancy, of the body's internal environment.

You may recall the S-shaped pattern from Figure 1.2 (A):

5, Order

Remember that this pattern indicates the maintenance of a relative constancy of oral body temperature within its normal range. This S-shaped pattern therefore also generally indicates the existence of homeostasis of any measured aspect of body structure or function. Claude Bernard, for instance, measured blood *glucose* (**GLOO**-kohs) concentration in dogs. Glucose is the most important sugar or "sweet" (*gluc*) substance within the bloodstream. It is the major fuel for our body cells.

Using the experimental method, Bernard found that blood glucose concentration (like oral body temperature) remains relatively constant within its normal range. Thus it, too, shows the characteristics of homeostasis.

Modern biology (especially physiology) owes much to Claude Bernard for demonstrating the importance of the experimental method in detecting homeostasis and other "truths" about living organisms.

Quiz

Refer to the text in this chapter if necessary. A good score is at least 8 correct answers out of these 10 questions. The answers are listed in the back of this book.

1. An organism is best described as:
    (a) A collection of body structures
    (b) A group of related body functions
    (c) Any highly organized thing
    (d) A living thing with a high degree of order

2. The structure of an organism is:
    (a) The entire program of function it carries out
    (b) The arrangement or pattern built up from its component parts
    (c) The highly disordered sum of all important body parts
    (d) The body parts which do not seem to fit together very well

3. Anatomy is:
    (a) Body structure and the study of body structures
    (b) Body function and the study of body functions
    (c) The study of both body structure and body function
    (d) Seldom considered whenever body structures are being studied

4. Anatomy differs from physiology in that:
    (a) Only physiology occurs in living things
    (b) Anatomy occurs in both living and dead organisms (not rotted away)
    (c) Physiology looks at the nature of body structures
    (d) Anatomy only occurs in dead things

5. Biological Order:
    (a) Explains the occurrence of disease and injury
    (b) Represents a particular pattern associated with living things
    (c) Does not really apply to the regulation of body temperature or blood sugar
    (d) Often prevents the body from responding to changes in the environment

6. An S-shaped curve followed over time generally indicates:
    (a) Lack of homeostasis
    (b) Biological Disorder in the external environment
    (c) A heavy intensity of environmental changes
    (d) The maintenance of some body structure or function within its normal range

7. Biological Disorder is usually tied to:
    (a) The creation of a particular pattern associated with living things
    (b) A healthy and disease-free state
    (c) Breaking of a particular pattern associated with living things
    (d) Normal body structure and function

8. Ecology is best defined as:
    (a) The removal of diseased and damaged plants and animals from the environment
    (b) Study of the relationships between different organisms and their environment
    (c) The wide occurrence of Biological Disorder within the external environment
    (d) A complete lack of association between different organisms and their environment

9. Natural History essentially approaches Biological Order by:
    (a) Naming and classifying newly discovered organisms
    (b) Testing a formal hypothesis
    (c) Applying the experimental method in a consistent manner
    (d) Conducting experiments on living organisms and reaching conclusions from the results

10. Homeostasis:
    (a) Does not usually exist in healthy individuals
    (b) Represents a "control of sameness" or relative constancy of the body's internal environment
    (c) Is basically the same thing as ecological relationships
    (d) Was extensively studied and defined by Aristotle

The Giraffe ORDER TABLE for Chapter 1
   (Key Text Facts About Biological Order Within An Organism)

1. _____
2. _____
3. _____
4. _____
5. _____

The Dead Giraffe DISORDER TABLE for Chapter 1
   (Key Text Facts About Biological Disorder Within An Organism)

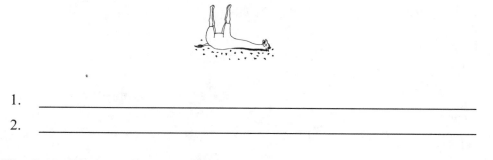

1. _____
2. _____

The Spider Web ORDER TABLE for Chapter 1
   (Key Text Facts About Biological Order Beyond the Individual Organism)

1. _____
2. _____

The Broken Spider Web DISORDER TABLE for Chapter 1
(Key Text Facts About Biological Disorder Beyond the Individual Organism)

1. _____

2

Patterns of Life

Chapter 1, **The Coming of Biology**, brought home to you some of the fundamental concepts of the modern "study of life." Prominent among these concepts was the notion of Biological Order – a particular pattern associated with living things. Recall that the spotted body pattern of a giraffe was used to symbolize Biological Order within the individual organism. And a spider in its web represented Biological Order extending outward beyond the organism. This elegantly simple visual pattern suggested a criss-crossing network of linkages existing among various living things in the external environment.

The Pyramid of Life (Levels of Biological Organization)

Now it is time to consider still another pattern, one well-known to the Ancient Egyptians. Of course, you have already guessed it! It is the distinctly pointed-and-sloped pattern of the pyramid:

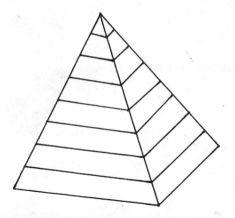

The pyramid shape consists of a number of horizontal levels, stacked one upon the other. This stacked pyramid is sometimes used in biology to help model the various *levels of biological organization.* A level of biological organization represents a certain degree of size and complexity of body structures, as well as the inter-relationships between them and other non-body structures. In biology, a *Pyramid of Life* can be identified. This Pyramid consists of 12 stacked horizontal levels of biological organization (Figure 2.1, A). Observe that these levels are numbered from bottom-to-top (Levels I–XII). Level XII, the peak of the Pyramid of Life, is the largest and most complex. This position at the top or peak symbolizes the fact that Level XII contains all the other levels of biological organization below and within it. Further, each of the other levels likewise contains the lower levels closer to the broad

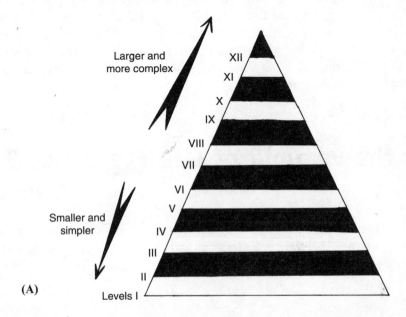

(B)

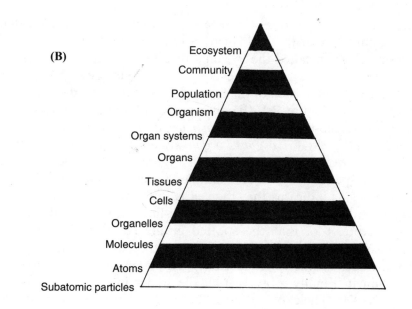

(C)

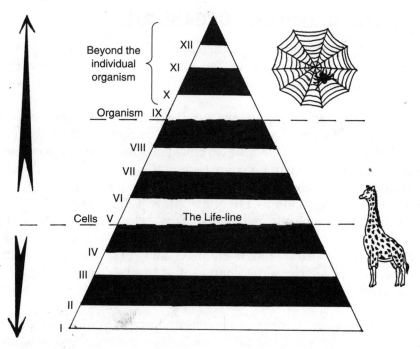

Fig. 2.1 The pyramid of life: The 12 levels of biological organization.

base of the Pyramid. The farther one goes up in the Pyramid, the greater is the size and complexity of the biological patterns encountered.

"Why is this diagram called the Pyramid of Life?" you may well ask. The reason is that life begins at a certain level of the Pyramid (Level V), and then continues upward through each of the higher levels of biological organization (Levels VI–XII). As Figure 2.1 (B) shows, the 12 Pyramid levels are as follows: *subatomic* (sub-ah-**TAH**-mik) *particles, atoms, molecules, organelles* (**OR**-gah-**nels**), *cells, tissues, organs, organ systems, organism, population, community*, and *ecosystem* (**E**-koh-**sis**-tem). Figure 2.1 (C) labels and tags certain levels in special ways. For example, Level V (cells) is labeled as "The Life-line." This indicates that the cell is the lowest *living* level of biological organization. All levels above this "Life-line," therefore, also include living things. Note, too, that the spotted giraffe (Biological Order) icon appears alongside Levels I–IX. This is because the giraffe, as an organism (Level IX), includes all the lower levels (organ systems down to subatomic particles) within it. Finally, observe that the spider in its web icon occurs beside Levels X–XII. The explanation is that a population, community, and ecosystem are all above the organism level of the Pyramid.

1, Order

1, Web

THE CHEMICAL LEVEL OF ORGANIZATION

Not specifically mentioned, yet, is the *chemical level* of biological organization. The chemical level includes the lowest three levels of the Pyramid (Figure 2.2). Specifically, this includes subatomic particles, atoms, and molecules.

The most important of these is the atom. An atom is the simplest form of a chemical *element*, or primary type of matter. A carbon (C) atom, for instance, is the simplest form of the element carbon. The four most common elements (atoms) found in the human body are carbon (C), oxygen (O), hydrogen (H), and nitrogen (N). [**Study suggestion:** Remember the four letters, COHN, as in the slang expression, "Don't COHN me, Man! Tell me the truth!"]

2, Order

The subatomic level is the one immediately "below" (*sub-*) the entire atom. This level consists of three main types of subatomic particles: *protons* (**PROH**-tahns), *neutrons* (**NEW**-trahns), and *electrons* (e-**LEK**-trahns). Each proton has an electrical charge of +1, while each neutron (as its name suggests) is electrically "neutral" (neither positively nor negatively charged).

Together, a certain number of protons and neutrons make up the *nucleus* (**NEW**-klee-**us**) or central "kernel" (*nucle*) of the atom. And orbiting very rapidly around this nucleus are one or more negatively-charged electrons.

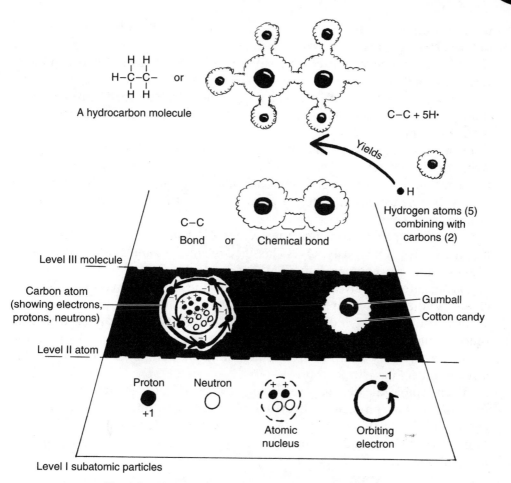

Fig. 2.2   The chemical level of biological organization.

The carbon atom, for example, contains 6 protons and 6 neutrons within its nucleus. Orbiting around it is a total of 6 electrons. Since the number of positively charged protons in the nucleus always equals the number of negatively charged electrons rapidly orbiting around the nucleus, the net charge of any atom is 0. Thus, the carbon atom, which has a nucleus of charge +6 (from the six protons), and a cloud of electrons with a charge of −6, has a net (overall) charge of 6−6 = 0. [**Study suggestion:** As shown in Figure 2.2, visualize the atom as being a hard gumball – the nucleus – surrounded by a sticky cloud of cotton candy, representing the movements of all the electrons orbiting around the nucleus.]

Lying just above the atom level is the molecule. A molecule is a combination of two or more atoms held together by *chemical bonds*. A chemical bond is a linkage created by the sharing or transfer of electrons between the outer surface clouds of atoms. For example, a C–C (carbon–carbon) bond (Figure 2.2) can be visualized as resulting from the linkage of two gumballs. Each gumball (nucleus of a C atom) is surrounded by a sticky (electron) cloud of cotton candy. When the two clouds are smashed together, and then pulled apart slightly, a chemical bond or sticky connection is made between them.

If one or more carbons and other types of atoms are bonded together in a string or ring shape, an *organic molecule* results. A very simple organic (carbon-containing) molecule is CO_2 – *carbon dioxide* (die-**OX**-eyed). This molecule is the one usually excreted as a waste product by the metabolism of our human body cells.

Our cells typically consume the oxygen (O_2) molecule in the process of making energy. Obviously, O_2 is classified as an *inorganic* molecule, because it does "not" (*in-*) contain any C atoms.

Carbon atoms easily bond together with one another, making long C–C chains. Quite often, hydrogen (H) atoms bond to the sides of these carbon atoms. The resulting molecules are called *hydrocarbons* (**HIGH**-droh-**kar**-buns) – literally, "hydrogen (*hydr*) and carbon" molecules. Most of the organic molecules making up human, animal, and plant bodies include large numbers of hydrocarbons.

When a molecule becomes very "big" (*macro-*), it is described as a *macromolecule* (**MAK**-roh-**mall**-uh-kyewl). One of the most important macromolecules in all living things is the *DNA molecule*. (Chapter 4 will provide much more information about chemical bonding, DNA, and other types of macromolecules essential for biology.)

ORGANELLES

Located just above the macromolecules (the top of the chemical level), are the organelles. The word, organelle, literally means "little organ." Hence, an organelle is a "little organ"-like structure that carries out specific functions within the cell. The cell, like an individual atom, contains a central "kernel" or nucleus. In the case of the cell, however, the nucleus is a rounded, kernel-like organelle. [**Study suggestion:** Picture an oval kernel of corn.] The cell nucleus of all humans, plants, and animals holds a large amount of the macromolecule, DNA. *Genes* (jeans) are particular sections of a DNA molecule. Their chief function is directing protein synthesis (the making of proteins) within the cell. (Chapter 5 will discuss all the major organelles and other components of the cell in more depth.)

Cells to Organism: Where the Pyramid Comes to Life

So far, we have been discussing the lowest levels of biological organization. Remember that the chemical level includes subatomic particles, atoms, and molecules. And just beyond the macromolecules lie the organelles. Finally, recall that the cell level begins "The Life-line," because it is the lowest *living* level of biological organization. It is now appropriate for us to ask, "Just what is it that we mean, by *living* level?" For an answer, just go back to Chapter 1, and review the section on the characteristics of living things. These are the characteristics that first appear in the Pyramid at the cell level, and then continue upward through the entire organism.

TISSUES

Right above the cell is the tissue level. A tissue is a collection of similar cells, plus the *intercellular* (**in**-ter-**SELL**-yew-lar) *material* located "between" them. There are four *basic or primary types of tissue* (Figure 2.3). These are *epithelial* (**eh**-pih-**THEE**-lee-al) *tissue, muscle tissue, connective tissue,* and *nervous tissue*. By basic or primary, it is meant that all of the specific types of tissues found in living things (especially humans and animals) are modifications or specializations of these four types.

3, Order

Epithelial denotes something present "upon" (*epi-*) the "nipples" (*theli*), such as the nipples in humans and related animals (like a bear, monkey, or your family cat or dog). Epithelial tissue is a covering and lining tissue. It covers body surfaces in general, and lines cavities within the body interior. Epithelial tissue, for example, forms the outermost layer of the human skin. As evident from Figure 2.3, A, epithelial tissue is almost entirely cellular in nature, with little or no intercellular material between its cells. Connective tissue (Figure 2.3, B), in great contrast, includes a lot of intercellular material between its cells. Frequently, this intercellular material contains long, slender rods – *connective tissue fibers*. Such fibers help connective tissue do its main job, which is to directly or indirectly connect body parts together. The fearsome teeth in the jaw of a shark, for instance, are firmly anchored into their sockets by connective tissue fibers. Nevertheless, these teeth sometimes break off, and then re-grow. Muscle tissue (Figure 2.3, C) consists of long, slender, *muscle fibers*. These muscle fibers are actually cells that contract or shorten, thereby creating body movements. Nervous tissue (Figure 2.3, D) is the

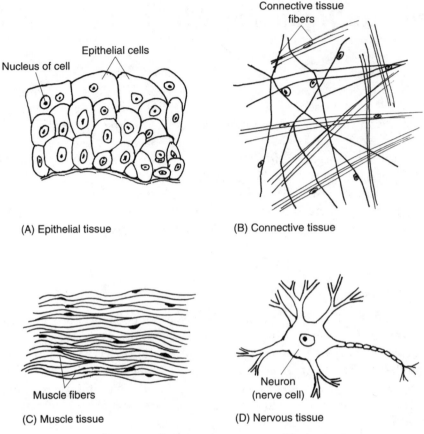

Connective tissue fibers

Epithelial cells

Nucleus of cell

(A) Epithelial tissue

(B) Connective tissue

Muscle fibers

(C) Muscle tissue

Neuron (nerve cell)

(D) Nervous tissue

Fig. 2.3   The basic or primary types of body tissues.

major tissue for communication and control within the body's internal environment. (The internal environment is everything deep in from the surface of the skin.) The nervous tissue largely does its communicating by means of *neurons* (**NUR**-ahns), the nerve cells. Neurons (nerve cells) within the nervous tissue inform the brain when the body has been damaged, usually resulting in the sensation of pain.

4, Order

ORGANS

In line above the tissues are the organs. An organ is a collection of two or more of the basic body tissues, which together perform some special function. The heart, for instance, is an organ with special muscle in its thick walls that allows it to carry out the function of pumping the blood. The heart contains connective tissue valves in its chambers, which are lined with modified epithelial cells. And various nerves supply the heart wall, serving either to speed it up or slow it down, whenever particular conditions arise.

ORGAN SYSTEMS

An organ system is a collection of two or more organs, which together perform some complex body function. Consider the *cardiovascular* (**kar**-dee-oh-**VA**-skew-lar) or *circulatory* (**SIR**-kyew-lah-**tor**-ee) *system*. (Examine Figure 2.4.) The cardiovascular (circulatory) system mainly involves the "heart" (*cardi*) and its attached blood "vessels" (*vascul*), through which the blood is pumped in a "little circle" (*circul*). The complex function performed by this system is the carrying of blood and its contained nutrients (such as oxygen and glucose) towards the body tissues, followed by the removal of waste products (such as carbon dioxide) from the body tissues.

5, Order

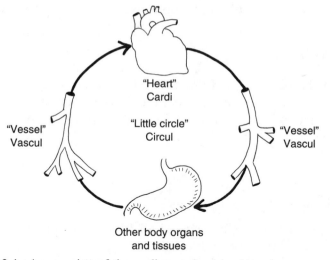

Fig. 2.4 An overview of the cardiovascular (circulatory) system.

Life beyond the Organism: Populations, Communities, and Ecosystems

Life does not stop with the individual body. There are still several levels of biological organization lying beyond the organism (but including it).

POPULATIONS

2, Web

Of course, male and female organisms of the same *species* (**SPEE**-sheez) or "kind" can breed with one another. The result of such breeding is a population. A population consists of a group of individuals of the same species that live together in the same place. A population of human beings, for example, consists of individuals of the species *Homo sapiens* (**HOH**-moh **SAY**-pea-**enz**) – literally "men" (*Homo*) who are "wise" (*sapiens*), living and breeding together.

COMMUNITIES

3, Web

Found above the population level is the community level. A community consists of all the organisms (including those of different species) living and potentially interacting with each other in the same area. An interacting group of *Homo sapiens* and various species of plants, animals, birds, and insects, for instance, could make up a particular forest community.

ECOSYSTEMS

4, Web

Perched at the peak of the Pyramid of Life is the ecosystem. An ecosystem includes both the living organisms in a biological community as well as the non-living factors present in their physical environment. A forest community like the one just described, for example, could include rivers, waterfalls, and scenic gorges.

SUMMARY

1, B-web

We have now defined and discussed all 12 levels of biological organization. And when we talk about biological organization, we essentially mean Biological Order. Each of the 12 levels in the Pyramid of Life, then, represents a particular pattern and complexity of Biological Order. If one of these levels becomes disturbed, such that its pattern is broken, then many other levels in the Pyramid may also become disturbed. Consequently, Biological Disorder may first occur within only a single level of biological organization. But this initial state of disorder may eventually disturb either higher or lower levels within the Pyramid. Therefore, Biological Disorder can rapidly spread through many levels, like a series of falling dominoes. The result – sickness or death of the organism! (Later chapters will explore some of these "falling dominoes," using specific examples.)

Homeostasis versus Ecological Relationships

Chapter 1 introduced the concepts of both homeostasis and ecology. Recall that homeostasis is a relative constancy of some particular aspect of the body's internal environment. The maintenance of a relative constancy of glucose level within the bloodstream was cited as a definite instance of homeostasis. Such conditions of homeostasis are especially important within the group of organisms called *mammals* (**MAM**-als) – animals with "breasts" (*mamm*) used to give milk to their young. Thus, cats, dogs, monkeys, and human beings are all mammals in which the blood glucose concentration is tightly regulated within a state of homeostasis. The glucose in the blood eventually becomes glucose within the breast milk, which in turn provides vital nutrition for the nursing youngsters.

The S-shaped pattern of homeostasis indicates a type of up-and-down balance within the body's internal environment. Homeostasis is therefore restricted to the organism level and below, within the Pyramid of Life.

Chapter 1 also discussed the idea of ecology – a "study of the household affairs" or relationships between different organisms and the total external environment. Such *ecological* (e-koh-**LAHJ**-ih-**kal**) *relationships* also represent a type of balance. But this balance exists at the levels of biological organization lying beyond the organism. That is, ecological relationships exist at the population, community, and ecosystem levels. Consider, for example, a so-called "population balance." This phrase represents the idea that, when the number of births occurring within a certain population equals the number of deaths in that population, over, say, a period of one year, then that population is "balanced" or "stable." It is neither greatly increasing nor greatly decreasing in size over time. This rough balance in the number of births versus deaths of particular organisms (such as humans) also greatly influences the surrounding environment. A stable human population essentially means a fairly stable consumption of food and water, as well as a stable excretion of waste products into the external environment.

5, Web

Quiz

Refer to the text in this chapter if necessary. A good score is at least 8 correct answers out of these 10 questions. The answers are listed in the back of this book.

1. A level of biological organization represents:
    (a) Some level of complexity below an organism
    (b) A particular layer within an Ancient Egyptian pyramid
    (c) A certain amount of size and complexity of body structures, along with the inter-relationships between them and various non-body structures
    (d) An almost complete lack of Biological Order

2. The lowest living level of biological organization is:
    (a) The organelle
    (b) The cell
    (c) Several types of subatomic particles
    (d) The ecosystem

3. The chemical level of organization includes:
    (a) Organelles, cells, and communities
    (b) Subatomic particles, atoms, and organs
    (c) Atoms, molecules, and subatomic particles
    (d) Every living level of biological organization

4. The cell nucleus represents the:
    (a) Organelle level
    (b) Tissue level
    (c) Organ system level
    (d) Molecule level

5. A molecule is a combination of two or more atoms held together by:
    (a) Chemical bonds
    (b) Genes
    (c) Hydrocarbon fragments
    (d) Electron–proton connections

6. A tissue is best defined as:
    (a) The smallest living level of biological organization
    (b) A collection of similar cells, plus the intercellular material between them

   (c) The intercellular material between cells, and not the cells
        themselves
   (d) A collection of similar cells, not including anything else

7. A collection of two or more organs, which together perform some
    complex body function:
    (a) Organism
    (b) Tissue
    (c) Organ system
    (d) Population

8. Snowshoe hares, bobcats, and arctic foxes all living in the same cold
    northerly area make up:
    (a) A population
    (b) A community
    (c) An organ system
    (d) An ecosystem

9. An ecosystem:
    (a) Exists at a level below the community
    (b) Only focuses upon the members of a particular population
    (c) Includes non-living factors in the physical environment
    (d) Never can be more complex than the organisms it contains

10. Ecological relationships are similar to homeostasis in that they:
    (a) Can always be diagrammed in an S-shaped pattern
    (b) Represent a relative constancy of the mammal's internal
         environment
    (c) Are restricted to the organism level and below
    (d) Maintain a rough balance or equilibrium over time

The Giraffe ORDER TABLE for Chapter 2
  (Key Text Facts About Biological Order Within An Organism)

1. _____
2. _____
3. _____
4. _____
5. _____

The Spider Web ORDER TABLE for Chapter 2
  (Key Text Facts About Biological Order Beyond the Individual Organism)

1. _____
2. _____
3. _____
4. _____
5. _____

The Broken Spider Web DISORDER TABLE for Chapter 2
  (Key Text Facts About Biological Disorder Beyond the Individual
Organism)

1. _____

3

Evolution: From Dawn to Darwin

Chapter 2 discussed The Pyramid of Life, and within it the various levels of biological organization. Recall that life begins at the cell level. It is now appropriate for us to ask, "Okay, we know what life is. But *where* did it come from? And did life forms *always* look the way they do on present-day Earth?"

Important Theories about the Origin of Life

Before life, and even before the planet Earth, the Universe or *Cosmos* (**KAHZ**-mohs) began as a "Big Bang." About 15 billion years ago, one or more huge, noisy "Bangs" (explosions) of dying stars created giant clouds of dust and gas. Out of such swirling clouds came our "sun" (sol) and the *solar* (**SOH**-lar) *system* of planets orbiting around it. The creation of our solar system was an important example of *Cosmic* (**KAHZ**-mik) *Order* – a pattern of organization that occurs within the Universe.

1, Web

The third planet from the sun – the Earth – appeared approximately 4.5 billion years ago. In the beginning, the Earth had a boiling-hot surface of liquid rock. But even at this early stage, carbon, hydrogen, and many other chemicals necessary for life, already existed.

As the planet cooled, a thin *crust* of hard rock formed upon the surface. Frequent eruptions of volcanoes forced gases out above this surface crust, thereby creating the *atmosphere*. The atmosphere is a 1-mile-thick blanket of gases surrounding the Earth. The early atmosphere contained water (H_2O) molecules in the form of water vapor, which was belched out by the volcanoes. Eventually, the surface cooled, and the water fell as rain. The rain pooled and filled vast surface canyons, giving rise to the first oceans. In addition, the cooling triggered the creation of oxygen (O_2) molecules.

Somewhere between 4 billion and 3.5 billion years ago, life probably began within the oceans. Biologists sometimes poetically call this great event the *Dawn of Life*. Figure 3.1 diagrams A TIME-CLOCK FOR ORDER. Notice that the CLOCK starts things off at midnight, representing 5 bya (billion years ago). As the hand of the CLOCK turns, the number of years from the present-day Earth decreases step-by-step from 5 bya. Each mark on the clock face represents a time span of 0.25 bya or 250 mya (million years ago). The numbers 4, 3, 2, and 1, consequently, represent 1 billion years ago, each. The appearance of the Earth, for instance, is marked about half-way between 5 and 4 (or 4.5 billion years ago). The Time-Clock sounded an alarm, in a sense, and woke up the Earth, at approximately 4 bya. The alarm rang out

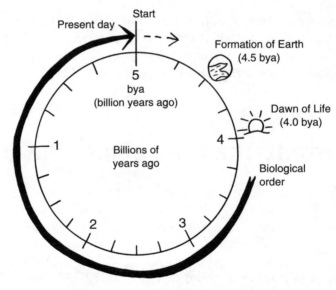

Fig. 3.1 A Time-Clock for order.

loudly, because it announced the Dawn of Life. And ever since this dramatic time, Biological Order has been present. It has continuously been reflected within the living things of our world.

1, Order

SPONTANEOUS GENERATION: LIFE COMING OUT OF NOWHERE

One of the theories attempting to explain the appearance of living things (and their associated Biological Order) is *spontaneous generation*. According to this theory, life is "produced" (generated) "automatically" (spontaneously) from non-living things. For many centuries, it was commonly assumed that a living fish, for instance, could arise "spontaneously" from the mud in the bottom of a lake! And people naively thought that rotten meat was somehow "spontaneously" transformed into swarms of living flies!

A scientist named *Francesco Redi* (**REH-**dee) threw doubt upon the Spontaneous Generation Theory in the 1600s. Redi demonstrated that when rotting meat is covered with a mesh, no flies come out of it. Yet, numerous people still believed that spontaneous generation produced the great variety of *micro-organisms* (**MY**-kroh-**or**-gan-izms) – "tiny organisms," such as bacteria.

Another scientist, *Louis Pasteur* (pas-**STER**), finally showed (in the mid-1800s) that micro-organisms did not spontaneously generate. Pasteur boiled a rich broth and kept it in a glass flask with a curved neck for well over a year. The curved neck prevented any micro-organisms from reaching the broth, so the broth remained clear and free of micro-organisms. If any micro-organisms had spontaneously generated within the broth, they would have turned the broth cloudy.

THE MODERN THEORY OF BIOGENESIS

After Pasteur's experiment, the old idea of spontaneous generation gave way to the modern *Theory of Biogenesis* (**buy**-oh-**JEH**-neh-sis). Biogenesis is the theory that "life" (bio) is "produced" (genesis) only from other things that are already alive. Two living male and female human beings, for instance, can mate and produce offspring that are also alive.

But our present-day observations about biogenesis still don't answer the question, "How did the very first life come into being on our planet?" Most evidence seems to point to a *primordial* (pry-**MORE**-dee-al) *soup* on the surface of the early Earth. Primordial means "first in order"; that is, existing first

2, Order

3, Order

before other things. The basic idea is that a primordial soup of complex organic compounds existed within the early oceans. Under the right conditions, such as the input of lots of energy from lightning striking the water and volcanoes spewing hot lava, these complex organic chemicals combined together and acquired a membrane around themselves. After this, they somehow became alive. Perhaps these first primitive micro-organisms were a species of ancient bacteria. According to the Theory of Biogenesis, these ancient bacteria eventually gave rise to all other living things on our planet.

Dinosaurs Help Make A Fossil Record

We will probably never know for sure exactly how life began – Why? The first living things (such as primitive bacteria) left behind no *fossil record*. Soft-bodied micro-organisms, such as the ancient bacteria, had no hard parts. Hence, their remains were rarely preserved.

Fossil means "to dig." Thus, the fossil record consists of the remains of ancient living things that have been preserved and "dug" up. With no hard body parts preserved, then, the first living creatures essentially left no fossils to be "dug" up later!

Nevertheless, many ancient soft-bodied bacteria created *stromatolites* (stroh-**MAT**-uh-**lights**) – "layered rocks." Even today, huge mounds of bacterial colonies live in shallow ocean water. Fine particles of dirt, calcium, and other minerals in the seawater, collect as sediment upon the colonies. This sediment eventually arranges into thin "layers" (*stromas*) streaking through each bacterial mound. As the layers or stromas become *calcified* (hardened with calcium sediment), they turn into layered rocks.

Figure 3.2 shows such a stromatolite. Among the oldest-known fossils are narrow, *filament* ("thread")-shaped bacteria, preserved within ancient stromatolite rock from Western Australia. They are about 3.5 billion years old.

THE FIRST LIVING ORGANISMS: TINY GREEN THREADS WITHOUT ANY "KERNELS"

In the living stromatolites found on present-day Earth, the bacteria are usually bluish-green in color. This green color indicates that they engage in *photosynthesis* (**FOH**-toe-**sin**-theh-sis). Photosynthesis is the use of "light" (*photo*) to "place (things) together" (*synthesis*). Specifically, photosynthesis is the process whereby certain types of organisms (usually green) use the energy in sunlight to

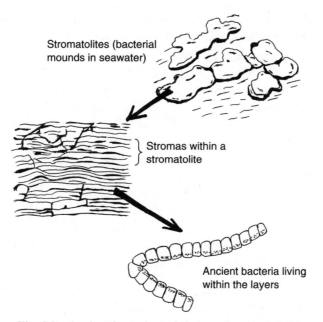

Stromatolites (bacterial mounds in seawater)

Stromas within a stromatolite

Ancient bacteria living within the layers

Fig. 3.2   Ancient bacteria and their rocky stromatolite.

make sugar molecules for themselves. Thus, the first ancient bacteria probably looked like tiny bluish-green threads or filaments. Further, each of these ancient, threadlike, green bacteria cells lacked a nucleus.

In Greek, the word part for "nucleus" or "kernel" is *kary* (**KAIR**-ee). Consequently, these ancient threadlike bacteria are often called *prokaryotes* (**proh-KAIR**-ee-oats). The reason for this name is that, being extremely ancient, the prokaryotes probably appeared "before" (*pro-*) other types of more advanced cells having a "nucleus" (*kary*).

CELLS WITH NUCLEI: IN COME THE "KERNELS"

For well over a billion-and-a-half years, it appears that the prokaryotes (photosynthetic cells without nuclei) were alone on this planet. All that time, however, these tiny green cells produced abundant amounts of oxygen (O_2) molecules, since oxygen is one of the main by-products of photosynthesis. The prokaryotes accomplished the critical task of making Earth's atmosphere capable of sustaining *aerobic* (air-**OH**-bik) *metabolism*. This is the type of metabolism that "pertains to" (*-ic*) "oxygen-or-air-using" (*aer*) "life" (*ob*).

The stage was thus set for the appearance of more complex cells, many of which had aerobic metabolisms dependent upon a ready supply of oxygen.

2, Web

About 2.1 billion years ago, numerous *eukaryotes* (**yew-KAIR**-ee-oats) evolved from the prokaryotes. Each eukaryote cell has a "good" (*eu-*) "nucleus" (*kary*) surrounded by its own membrane. In addition, the eukaryote cell contains numerous other organelles, each enclosed within its own individual membrane. Besides the nucleus, another very important membrane-covered organelle is the *mitochondrion* (**my**-toe-**KAHN**-dree-un). The mitochondrion carries out most of the aerobic (oxygen-using) metabolism within the eukaryote cell.

Figure 3.3 provides a summary of the above information. It shows that there are two basic types of cells – prokaryotes (cells without a nucleus) and eukaryotes (cells having a nucleus and other organelles surrounded by membranes). Further, we see that the major examples of prokaryotes now in existence are the bacteria. The cells of nearly all other organisms besides bacteria, in contrast, are eukaryotes. (A detailed discussion of cell anatomy and physiology will be provided in Chapter 5.)

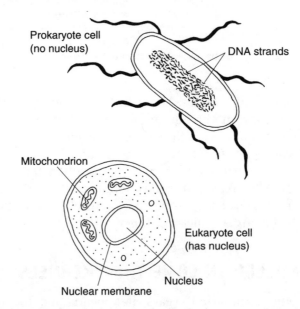

Fig. 3.3   The two basic types of cells: Prokaryotes versus eukaryotes.

UNICELLULAR ORGANISMS EVOLVE INTO MULTICELLULAR ONES

So far, we have traced the development of life up to 2.1 billion years ago. Both the ancient prokaryotes and eukaryotes had a critical characteristic in common. They were *unicellular* (**YEW**-nih-**sell**-yew-lar) or "single" (*uni-*) "celled" (*cellul*) organisms.

Around 1.5 billion years ago, the first *multicellular* (**MUL**-tih-**sell**-yew-lar) or "many-celled" organisms appeared. This event was very important, because before this time all living organisms were *generalists*. Each unicellular organism carried out all of the general processes of life. But with the coming of multicellular creatures, *specialization* of body structure and function arrived. With many cells available, some of the cells could do one body function, while other groups of cells could become specialized to carry out other body functions.

4, Order

Another fact about multicellular organisms is that they usually develop from a single fertilized *ovum* (**OH**-vum), or "egg" cell. A multicellular organism results when the fertilized ovum undergoes repeated cell division and *differentiation* (**dif**-er-en-she-**AY**-shun) – "a process of becoming different" or specialized.

The *multicellular eukaryote organisms*, then, are the many-celled creatures whose cells contain *nuclei* (**NEW**-klee-eye) or "kernels." The very first such organisms appearing in the fossil record are relatively small *algae* (**AL**-jee) or "seaweeds," 1.5 billion years ago. These algae were greenish in color and engaged in photosynthesis. Algae are considered the most primitive members of the Plant Kingdom. They have no roots, stems, or leaves. Modern multicellular algae can form huge floating seaweeds, however.

A *plant* or "sprout" is an organism that contains the substance *chlorophyll* (**KLOH**-roh-**fill**) and carries out photosynthesis for creating its energy. Chlorophyll literally means "green" (*chlor*) "leaf" (*phyll*). It is the green pigment found in leaves and other parts of plants. Chlorophyll absorbs sunlight, thereby providing the energy used by photosynthesis to make sugars for plant cells.

Paleobotanists (**PAY**-lee-oh-**BAHT**-uh-nists) are "those who specialize in" (*-ist*) "ancient" (*paleo*) "plants" (*botan*). When paleobotanists studied the fossil record, they concluded that land plants (like trees, ferns, and flowers) evolved from primitive multicellular algae, about 500 million years ago.

Related to the plants are the *fungi* (**FUN**-jeye). A *fungus* (**FUN**-gus) is a plant-like organism that acts as a parasite, living on dead or living organic matter. Perhaps you immediately picture a "mushroom," after which this group is named. The mushrooms often feed on decaying leaves. Fungi can be either unicellular (as in yeast cells), or multicellular (as in molds and mushrooms). Fungi contain no chlorophyll, so they cannot utilize photosynthesis. Many fungi are *pathogenic* (**PATH**-oh-**jen**-ik) or "disease" (*path*) "producing" (*gen*) for both plants and animals. Common examples are the

1, B-Web

5, Order

occurrence of white spots of mildew fungus on damp leaves, and the existence of yeast infections in the female vagina.

The first fossils of fungi are recorded at about the same time as those of land plants (460–500 million years ago). One can reasonably speculate that the first fungi were probably pathogenic parasites clinging to the moist leaves of the first land plants.

"Well, what about the animals? What about the dinosaurs?" you may now be asking yourself. An *animal* is any "living, breathing" (*anima*) multicellular organism that is not a plant or fungus. Animals have eukaryote cells with nuclei, and they must eat other organisms or organic matter in order to survive. (We will learn much more about animals in later chapters.)

The oldest animal fossils are around 600 million years old. These first animals were multicellular ocean-dwellers, such as jellyfish, corals, and sea-worms. They were all *invertebrates* (in-**VER**-tuh-brits) – animals "without" (*in-*) spines or "backbones" (*vertebr*). Delicate jellyfish and other invertebrate animals gracefully floated in the sea during what is technically called the *late Pre-Cambrian* (pree-**KAM**-bree-un) *Era*.

About 500 million years ago, some animals became *vertebrates*, developing a backbone. Among the first such vertebrates were jawless fishes. One hundred million years later, many animals joined the plants and fungi in coming out of the water to live on land.

This general time span (from about 500–200 million years ago) is called the *Paleozoic* (**pay**-lee-uh-**ZOH**-ik) or "ancient life" *Era*. This period saw the development of the first vertebrates (fishes, amphibians, and reptiles), land plants, insects, and vast forests of fern-like trees. *Amphibians* (am-**FIB**-ee-uns) literally "live a double life," meaning that they can occupy "both" (*amphi-*) land and water. Amphibians have a moist skin without any scales. They include frogs, toads, newts, and salamanders. Giant amphibians roamed as Kings of the Earth, long before the dinosaurs!

Reptiles derive their name from the Latin for "crawlers." Most groups of reptiles (such as turtles, lizards, alligators, and crocodiles) do a lot of creeping and crawling on land. We all know that snakes are a group of reptiles that slither, however. Reptiles breathe through lungs and usually have skin that is covered by either horny plates or flat scales. The reptiles evolved from the amphibians and appeared during the middle-to-late Paleozoic Era. But they didn't really become the dominant life form on Earth until the *Mesozoic* (**mess**-uh-**ZOH**-ik) or "middle" (*meso-*) "life" (*zo*) *Era*.

The Mesozoic Era is very vivid in people's minds, because it is often nick-named the *Age of Reptiles*. This period is literally "in the middle," because it began about 200 million years ago, just after the Paleozoic Era, with its "ancient life," and ended about 65 million years ago, just before the modern period we find ourselves in, today. The Mesozoic Era is very critical for the fossil record. The preserved bones suggest that flying reptiles likely evolved into birds, and small, shrew-like mammals arrived. Flowering plants bloomed in the forests and prairies.

But it is the *dinosaurs*, or "terrible" (*dino*) "lizards" (*saurs*), that completely dominated the Mesozoic Era. The dinosaurs were reptiles that lived on the land, whereas the *pterosaurs* (**TER**-uh-sors) – "winged lizards" – were flying reptiles that took command of the sky. Mention fossils, and most people probably visualize dinosaur bones! Small wonder, since some dinosaurs were the biggest creatures ever to walk the Earth! Recent evidence suggests that particular types of dinosaurs, such as the duck-billed dinosaurs, traveled in social groups and even cared for their young, after they hatched from eggs (see Figure 3.4).

3, Web

Fig. 3.4 "Mother love" and the duck-billed dinosaur.

"If the dinosaurs were so powerful, and some cared for their young, then why did they become extinct?" a curious person might ask. About 65 million years ago, Earth was nearing the end of the Mesozoic Era. According to the *impact hypothesis*, about this time a huge comet or asteroid crashed into the Earth at great speed. The force of impact created a massive cloud of dust and debris. This great cloud blocked most of the sunlight, killing plants and dramatically cooling Earth's tropical climate. Plant-eating dinosaurs had nothing to eat, and died out. Thus, meat-eating dinosaurs, which could not devour plant-eating dinosaurs, became extinct as well.

2, B-Web

The mass extinction of all the dinosaurs led to the *Cenozoic* (**sen**-uh-**ZOH**-ik) or "new" (*ceno*) "life" (*zo*) *Era*. The Cenozoic Era is sometimes nicknamed *The Age of Mammals*. The reason is because during this past 65 million years or so, leading up to the present day, mammals have become the dominant animals on the Earth. Mammals, like birds, probably evolved from reptiles. A *mammal* is an animal that nurses its young with its "breasts" (*mamma*). In addition, mammals are covered with hair and are *endothermic* (**en**-doh-**THER**-mik). Endothermic organisms have "inner" (*endo-*) control of their body "heat" (*therm*) or temperature. Therefore, early mammals had a much greater ability to adapt to the colder climate of the modern era, compared to reptiles and amphibians. They could find more *habitat* (**HAB**-uh-tat) on the land and water, a wider variety of colder places to "live in" (*habit*). [**Study suggestion:** Would you expect to see an alligator – a reptile – swimming in the freezing water of the Arctic Ocean? Or, would you be more likely to see a killer whale – a marine mammal – swimming there?]

Most of the main groups of mammals were in existence about 50–60 million years ago. One of these groups was the *primates* (**PRY**-mates). Since human beings (Homo sapiens) belong to the group of primates, it is only natural that we should label ourselves as being "of first rank or importance" (*primat*). Primates include monkeys, apes, and various other creatures, as well as humans. The apelike ancestors that may have evolved into humans appeared approximately 5 million years ago. Our species, Homo sapiens, finally showed up just 100,000 to 200,000 years ago. This time has often been nicknamed *The Ice Age*. This is due to the fact that fossils of early humans were found amid evidence of huge glaciers in Europe.

FOSSIL RECORD SUMMARY

4, Web

Table 3.1 provides a brief summary of the major *Geological* (**jee**-uh-**LAHJ**-uh-kul) or "pertaining to Earth-study" *Eras*. Observe from the table that there are four Eras: the Pre-Cambrian, Paleozoic, Mesozoic, and Cenozoic Eras. Not shown in the table are a number of *Periods*, into which each of the eras is subdivided. Finally, the most recent (Cenozoic) Era has two periods, each of which is subdivided into a number of *Epochs* (**EP**-uks). Since this book just provides an overview, the specific periods and epochs have not been identified. [**Study suggestion:** Picture three steps of geological time going down, big to smaller to smallest. The top step is the Era. The middle step is the Period. And the third step is the Epoch.]

The relative amount of time covered in each of the four major eras is easily visualized within the FOSSIL RECORD CLOCK (Figure 3.5). Recall that each mark on this CLOCK represents a time span of 0.25 bya (billion years ago), or 250 mya (million years ago). Therefore, the appearance of human beings a mere 100,000 years ago is so recent in Earth's overall history, that it occurred just the last second or so before our model CLOCK strikes its end at midnight!

Table 3.1   The four major eras and their characteristics.

Era	Time span	Major events
Pre-Cambrian Era	4.5 billion to 545 million years ago	Formation of Earth Primordial soup Stromatolites containing bacteria Unicellular prokaryotes, then eukaryotes Multicellular eukaryote organisms (algae) appear
Paleozoic Era	545 million to 245 million years ago	"Ancient Life" Era First vertebrates (fish), land animals (amphibians, reptiles), land plants, fungi, insects
Mesozoic Era	245 million to 65 million years ago	"Middle Life" Era, Age of Reptiles Flying reptiles become birds Small mammals and flowering plants appear Rise and fall of the dinosaurs
Cenozoic Era	65 million years ago to present day	"New Life" Era, Age of Mammals Modern mammals and birds evolve from reptiles Primates (monkeys, apes, apelike human ancestors) The Ice Age, including arrival of Homo sapiens

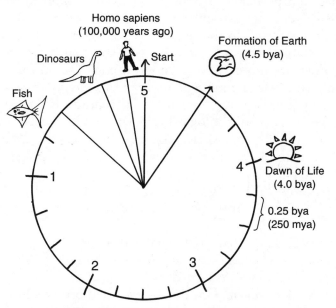

Fig. 3.5   The fossil record clock.

Charles Darwin and the Theory of Evolution

"How did all these different types of living organisms come to replace or join each other over time? What complicated process seems to be responsible?" you might well ask. The biological answers are: *evolution* and the process called *natural selection*.

Evolution means "a process of rolling out," a gradual process of developing or unfolding. In a biological sense, evolution can be considered a gradual unfolding or development of particular patterns of Biological Order, over long periods of time. These patterns of Biological Order are reflected in the body structures and functions of particular living organisms. "Okay, I see," you may persist, "But what process determines which patterns of Biological Order survive, which patterns become modified, or which patterns (like that of the dinosaurs) become extinct?

Enter *Charles Darwin*, an English scientist who stated his *Theory of Evolution by Natural Selection*. Darwin wrote an important book, *The Origin of Species by Natural Selection* (1859), that explained the fundamental ideas behind this theory. For five years (1831–1836), Darwin sailed around the world on a ship called the *Beagle*. He spent countless hours collecting fossils and living creatures along the coastline of South America, and in observing strange creatures only found in a set of tiny Galapagos Islands. He concluded that particular groups of organisms *adapted* themselves (over long periods of time) to particular types of habitats. Since habitats widely differed, he reasoned, so do the organisms that have adapted themselves to these different habitats. By *adaptation*, it is meant that particular patterns of Biological Order are more "fit" or "appropriate" for certain habitats than are other patterns. And the individual organisms having these more "fit" or "appropriate" patterns were more likely to reproduce themselves. Hence, their particular patterns of Biological Order would be passed on to succeeding generations of offspring. This produces, in a sense, a "survival of the fittest," or a "natural selection." The "fittest" (most efficiently adapted) organisms to a particular habitat are the ones that tend to survive and pass their patterns of Biological Order on to their offspring. Only the "most fit" are "selected naturally" to survive and reproduce within a certain habitat.

Charles Darwin supported his theory with many commonsense observations. For example, he cited the seasonal color changes in the feathers of the *ptarmigan* (**TAR**-muh-gun), a type of wild grouse found in cold and mountainous regions. Most ptarmigans he observed had brown spotted feathers in the summer, followed by a dramatic change in color pattern to all-white

5, Web

feathers during the winter. However, a few individual birds did not change their brown spotted feathers to white in wintertime. They were soon "naturally selected against" – easily seen and eaten by hungry bobcats or other predators roaming the frozen white landscape.

Darwin expanded this natural selection idea to explain the evolution or extinction of entire species of organisms. Evidence is still being accumulated to support this general concept, even today.

Quiz

Refer to the text in this chapter if necessary. A good score is at least 8 correct answers out of these 10 questions. The answers are listed in the back of this book.

1. Cosmic Order is best described as:
   (a) Just another version of Biological Order
   (b) A pattern of organization that occurs within the Universe
   (c) Lack of any apparent organizing framework throughout the Cosmos
   (d) Giant clouds of dust and gas in outer space

2. The Dawn of Life probably:
   (a) Occurred before the surface of Planet Earth began to cool and rain fell
   (b) Prevented evolution from progressing any farther
   (c) Started Biological Order as we know it
   (d) Happened only 10 million years ago

3. A heavy rain falls. Overnight, a dry, parched field becomes a swampy area behind your house. The next night, you hear thousands of frogs loudly croaking. If you lived before the time of the scientist Francesco Redi, you would likely conclude that:
   (a) Alien frogs dropped out of the sky during the rainstorm!
   (b) The frogs appeared suddenly in the usually dry field by spontaneous generation
   (c) The frogs mutated from micro-organisms
   (d) "Survival of the fittest" was at work

4. Biogenesis is the theory that:
   (a) Living organisms are only produced from other living organisms
   (b) Dead animals can give rise to living ones

(c) Sometimes plants can change into animals, and vice versa

(d) Life first arose from a primordial soup

5. "How did the very first life come into being on our planet?" According to the chapter, your answer to this question would be that:
   (a) Ancient dinosaurs changed back into primitive micro-organisms
   (b) The Ice Age froze everything solid, but then living things thawed out
   (c) Natural sources of energy (such as lightning and hot lava) sparked a fundamental change in the primordial soup of complex organic compounds within ancient oceans
   (d) Boiling hot lava fields changed automatically into primitive bacteria cells

6. These help provide the first solid evidence of life within the fossil record:
   (a) Stromatolites
   (b) Phagocytes
   (c) Fine particles of sediment
   (d) The bones of jawless fishes

7. Multicellular organisms:
   (a) Probably evolved into unicellular ones
   (b) Are generalists for the processes of life
   (c) Were the likely descendants of unicellular organisms
   (d) Only occasionally contain eukaryote cells

8. According to the impact hypothesis, the dinosaurs became extinct because:
   (a) Jungles took over too much of the planet, thereby destroying dinosaur habitat
   (b) The Earth cooled and dried out, due to a change in the tilt of its axis
   (c) A huge comet or asteroid hit the Earth with such force that it created a massive cloud of dust and debris in the atmosphere
   (d) The heat produced by the impact of a fiery comet melted both the North and South Poles

9. The Cenozoic Era:
   (a) Is often nicknamed "The Age of Giant Amphibians"
   (b) Has alternatively been called "The Age of the Dinosaurs"
   (c) Occurred when the mammals replaced the dinosaurs as the dominant organisms on our planet
   (d) Was finished long before human beings walked the Earth

10.   According to Charles Darwin's version of the Theory of Evolution:
   (a)   The fossil record strongly suggests that simpler organisms
         appeared after larger, more complex organisms
   (b)   Species competing for the same available resources in the
         environment went through a process of natural selection over time
   (c)   The strange animals of the Galapagos Islands were not fully
         adapted to their ecosystem
   (d)   A particular pattern of Biological Order has nothing whatsoever to
         do with the successful adaptation of a particular species to its
         environment

The Giraffe ORDER TABLE for Chapter 3
   (Key Text Facts About Biological Order Within An Organism)

1.   _____
2.   _____
3.   _____
4.   _____
5.   _____

The Spider Web ORDER TABLE for Chapter 3
   (Key Text Facts About Biological Order Beyond the Individual Organism)

1.   _____
2.   _____
3.   _____
4.   _____
5.   _____

The Broken Spider Web DISORDER TABLE for Chapter 3
(Key Text Facts About Biological Disorder Beyond the Individual Organism)

1. _____

2. _____

Test: Part 1

DO NOT REFER TO THE TEXT WHEN TAKING THIS TEST. A good score is at least 18 (out of 25 questions) correct. Answers are in the back of the book. It's best to have a friend check your score the first time, so you won't memorize the answers if you want to take the test again.

1. Which of the following terms translates exactly to mean the "study of life"?
    (a) Anatomy
    (b) Physiology
    (c) Ecology
    (d) Biology
    (e) Structure

2. All organisms have this in common:
    (a) Anatomy, physiology, and Biological Order
    (b) Anatomy only
    (c) Physiology only
    (d) Just the fact that they are alive
    (e) Living, but lacking any Pattern of Order

3. A young boy squashes a frog flat like a pancake under his boot. The dead frog no longer has:

(a) Biological Disorder
(b) Physiology
(c) Gross anatomy
(d) Microscopic anatomy
(e) Either structure or function

4. A severely fractured wing no longer permits a wounded bird to fly. The best theoretical explanation for this fact is:
   (a) The germ theory of disease
   (b) The state of Biological Disorder is too great for normal function to occur
   (c) Bird metabolism has been totally destroyed
   (d) A perfect pattern of body structure exists
   (e) Bird physiology has not been significantly affected

5. Biological Disorder is best reflected by:
   (a) Normal healing of a skin wound
   (b) Two diseases making a patient sicker
   (c) Sexual reproduction
   (d) A small bone growing to become a larger one
   (e) Extensive patterns of veins within a leaf

6. According to the text, a broken spider web could best be used as a symbol for:
   (a) Normal anatomy and physiology
   (b) A good balance among the organisms inhabiting a local swamp
   (c) Draining of a swamp, accompanied by massive loss of species and habitat
   (d) A fish dying from a swallowed hook
   (e) A football player falling and injuring his knee

7. The Father of Natural History:
   (a) Perry Como
   (b) Claude Bernard
   (c) William Tell
   (d) Aristotle
   (e) Hippocrates

8. Derives from the Latin for "control of sameness" (relative constancy):
   (a) Ecology
   (b) Glucose
   (c) Homeostasis
   (d) Excretion
   (e) Oral

9. Carbon (C) represents what specific level of biological organization?
   (a) Chemical bond
   (b) Molecule
   (c) Atom (element)
   (d) Electron
   (e) Proton

10. Which of the following is an example of an organic molecule?
    (a) C–C (carbon–carbon) bond
    (b) CO_2
    (c) O_2
    (d) N_2
    (e) NH_2

11. Level lying just below the cell, but just above the molecule:
    (a) Organism
    (b) Atom
    (c) Organelle
    (d) Tissue
    (e) Organ system

12. A "little-organ"-like structure that conducts specific functions within
    the cell:
    (a) Organelle
    (b) Population
    (c) Community
    (d) Organic chemical
    (e) Organism

13. A liver cell is alive, but its mitochondria technically are not, because:
    (a) Mitochondria have all the properties of living things
    (b) The cell is able to reproduce itself
    (c) The mitochondrion produces lots of energy
    (d) There can be no change in the number of mitochondria within each
        cell
    (e) Cells (but not mitochondria) are commonly visible through the
        compound microscope

14. Which of the following groups represent the basic (primary) types of
    tissue?
    (a) Cardiac, nervous, endocrine, muscular
    (b) Bony, connective, cartilage, muscle
    (c) Blood, epithelial, nervous, exocrine

(d) Brain, circulatory, nervous, cardiac

(e) Epithelial, connective, nervous, muscle

15. The kidney is considered an example of what level of biological organization?
    (a) Tissue
    (b) Cell
    (c) Community
    (d) Organ system
    (e) Organ

16. Male and female members of the Homo sapiens species are considered:
    (a) An ecosystem
    (b) A population
    (c) A community
    (d) A genus
    (e) A subphylum

17. Ecological relationships exist at what level of biological organization?
    (a) Organ, organ system, organism
    (b) Organ system, community, population
    (c) Population, community, ecosystem
    (d) Tissue, chemical, organelle
    (e) Organism, community, ecosystem

18. A swarm of mosquitoes is blown into a new swampland by strong winds. There are few birds to eat them. Soon the mosquitoes greatly multiply in number. This would be an instance of:
    (a) Population balance
    (b) Species extinction
    (c) Predator living upon prey
    (d) An unstable population
    (e) Homeostasis of mosquito numbers

19. Louis Pasteur demonstrated that:
    (a) The Theory of Biogenesis is false
    (b) Micro-organisms did not spontaneously generate themselves
    (c) Francesco Redi was very wrong about his finding that flies do not arise from rotting meat
    (d) Human beings came into existence during the "Big Bang"
    (e) The Ancient Egyptians did not arise by spontaneous generation

20. "Life is produced only from other things that are already alive." This statement summarizes the:
    (a) Concept of evolution by natural selection
    (b) Folklore about mice spontaneously generating from dirty shirts
    (c) Theory of Biogenesis
    (d) Doctrine of "Survival of the Fittest"
    (e) "One bird in the hand is better than two birds in the bush!"

21. Eukaryotes are characterized by:
    (a) The lack of a cell nucleus
    (b) A rigid cell wall, but no nuclear membrane
    (c) The presence of several membrane-surrounded organelles, including a cell nucleus
    (d) The complete absence of any organelles when viewed under the microscope
    (e) The same body structures and functions as most bacteria

22. With the coming of multicellular organisms, what features arrived?
    (a) A generalized ability for all body cells to carry out all of the essential functions of life
    (b) A rapid loss of physiology, accompanied by a corresponding gain in anatomy
    (c) Greater specialization of body structures and functions
    (d) Nothing really different than was present for the unicellular organisms
    (e) Each cell had two nuclei, rather than one

23. The Paleozoic Era is important due to the fact that:
    (a) It represents the Age of Reptiles
    (b) Eskimos first showed up during its Ice Age
    (c) The first vertebrate animals and fish, land plants, insects, and forests of fern-like trees, appeared during this Era
    (d) Invertebrate animals (like starfish and sea squirts) were the only ones around
    (e) Delicate jellyfish first appeared within the sea

24. "The Age of the Dinosaurs":
    (a) Cenozoic Era
    (b) Mesozoic Era
    (c) Late Pre-Cambrian Era
    (d) Paleozoic Era
    (e) The Final Ice Age

25. The last major group of mammals to evolve:
     (a) Birds
     (b) Snakes
     (c) Lizards and turtles
     (d) Fur-bearing marsupials
     (e) Primates

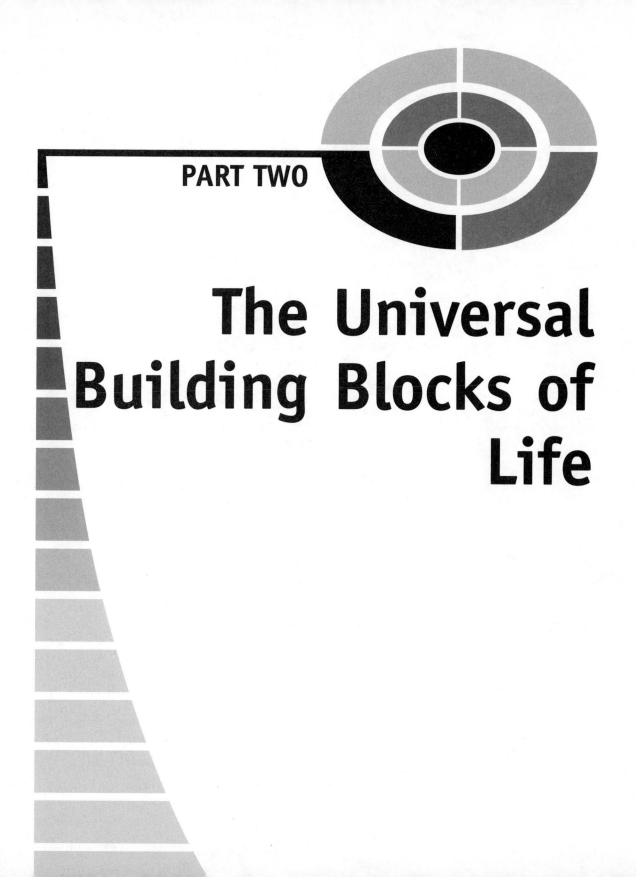

PART TWO

The Universal Building Blocks of Life

4

Chemicals: The Tiniest Blocks

In the first three chapters of this book, we were getting ourselves ready for biology. We painted with a wide brush, making broad strokes through several key areas of the subject matter that we will now examine in greater depth. First among these will be the chemicals – the tiniest blocks (or levels) in the Great Pyramid of Life.

Chemical Bonds: Builders of Order

Chapter 2 defined a chemical bond as the sharing or transfer of electrons between the outer surface clouds of atoms. And Figure 2.2 depicted a C–C (carbon–carbon) bond. The carbon atom nuclei were represented as two gumballs, each surrounded by a sticky electron cloud of cotton candy. The bond resulted when the clouds were smashed together, then slightly pulled apart.

1, Order

BONDS AND ORDER

Picture in your mind groups of three or more atoms held together by chemical bonds. Do you see triangles or other simple geometric patterns? In a functional sense, chemical bonds are builders of order. They are builders of order or pattern at the chemical level of biological organization. Consider two specific examples – a water (H_2O) molecule (Figure 4.1) and a sodium chloride (NaCl) crystal.

Fig. 4.1   The water molecule and its bonds.

The water molecule is shaped like a triangle or V, with the O (oxygen) atom at the tip or apex. A hydrogen (H) atom is attached on each side by a *covalent* (koh-**VALE**-ent) *bond*, one which involves a sharing of outermost electrons between atoms. Since the larger oxygen atom draws each of the electrons from the hydrogen atoms into closer orbit around itself, positive and negative *poles* are created. The O end is the *negative pole* (having an excess of two negatively charged electrons), while each H end makes up a *positive pole*. The H ends of each bond create a positively charged pole, because their electron has been unequally shared with the oxygen. In effect, the oxygen atom is an "electron hog"! It still leaves each electron-robbed hydrogen atom with its positively charged proton, however.

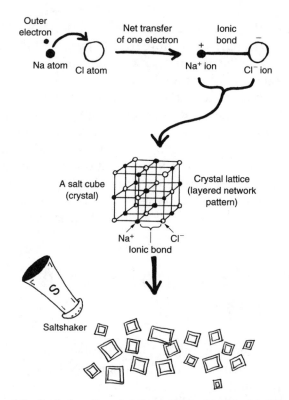

Fig. 4.2   Ionic bonding and the sodium chloride (NaCl) crystal.

Now let us consider the salt crystal (Figure 4.2). Numerous sodium (Na) and chlorine (Cl) atoms combine with one another to form a beautiful cube-shaped solid. This precise geometric pattern is seen in common table salt. (Think of this when you are eating your salted pretzels or French fries!) Each face of the salt cube is a square, with alternating Na^+ and Cl^- *ions* (**EYE**-ahns) at its corners. An ion is an atom that has either a net excess or deficiency of outermost electrons, such that it is electrically charged. The sodium ion, for instance, is symbolized as Na^+, because it has a deficiency of one electron. Where did this electron go? It was transferred from the outer cloud of the Na atom, and onto the cloud of the Cl atom. The result is an *ionic bond* – a bond that results from the transfer of one or more electrons between atoms. And since the *chlorine atom* (Cl) picks up that extra electron being transferred from sodium, it becomes a *chloride ion*, Cl^-, having an overall negative charge.

Thus we see that there are two main types of chemical bonds – covalent bonds and ionic bonds. Covalent bonds result from the sharing (either equal or unequal) of outermost electrons between atoms, while ionic bonds arise from the complete transfer of one or more outermost electrons between atoms. In either case, the result is an increase in the state of order or pattern of the atoms involved. In particular, note from Figure 4.2 (above) the elegant crystal *lattice* (**LAT**-is) or layered network pattern of the NaCl cube.

Recall from Chapter 2 that carbon dioxide (CO_2) is considered an organic molecule, because it contains a carbon (C) atom. Conversely, both the H_2O molecule and the NaCl crystal are inorganic compounds, because they do "not" (in-) contain any carbon. Long chains of hydrocarbon molecules contain numerous C–H (carbon–hydrogen) covalent bonds. Due to the fact that each carbon atom can form covalent bonds with other carbon atoms (as well as with H atoms), it can create extremely long, highly orderly, C–C chains. Such long organic chains form the stable "backbone" of many macromolecules (like DNA and proteins), as well as that of larger body structures.

Water and the Breakdown of Salt: A "Dissolving" of Order

Here is a chemical rule-of-thumb: "Like dissolves like." This statement means that chemicals which are "like" one another (in terms of their electrical charge and bonding) tend to dissolve or mix well with each other. Relevant terms here are *solvent* (**SAHL**-vent) versus *solute* (**SAHL**-yoot). A solvent is "one that dissolves," while a solute is the "thing being dissolved." The result of their mixture is a *solution*:

SOLUTE	+	SOLVENT	=	A SOLUTION
(thing dissolved)		(the dissolver)		(solute plus solvent mixed together)
Example : NaCl		H_2O		A saltwater solution

The most common solute within human and most animal bodies is sodium chloride (NaCl). Yes, sodium chloride is an inorganic crystal, but it is also a type of *electrolyte* (ee-**LEK**-troh-**light**). An electrolyte is a substance that "breaks down" (*lyt*) into ions when placed into water solvent, such that the resulting solution can conduct an "electrical current" (*electro-*). Sodium chloride, for example, has its Na^+ and Cl^- ions tightly bonded together within its crystal salt cube structure. And water has its O^{2-} and two H^+ atoms connected by polar covalent bonds. Note that both NaCl and H_2O are "alike" in that they contain net positive (+) and negative (–) charges within their structure. Hence, according to our chemical rule-of-thumb ("Like dissolves like"), water is an excellent "dissolver" or solvent for salt. The + charges of Na^+ are attracted to the negative (–) charges on the O^{2-} of many H_2O molecules. In contrast, the negative (−) charges of Cl^- are attracted to the positive (+) charges of the H^+ poles of the water molecules. Thus, sodium and chloride tear apart, making NaCl a solute that readily "breaks down" when placed into water solvent (see Figure 4.3).

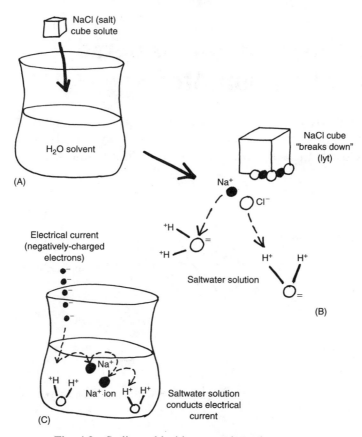

NaCl (salt) cube solute

H_2O solvent

(A)

NaCl cube "breaks down" (lyt)

Na^+

Cl^-

^+H
^+H
Saltwater solution

H^+ H^+

(B)

Electrical current (negatively-charged electrons)

Na^+
^+H H^+ Na^+ ion H^+ H^+
Saltwater solution conducts electrical current

(C)

Fig. 4.3   Sodium chloride as an electrolyte.

Now electricity (electrical current) is basically a flowing of negatively charged electrons. The saltwater solution resulting from the mixture of sodium chloride and water can therefore conduct this electrical current, because the negative (−) charges on the flowing electrons are attracted to many positive charges on the Na^+ ions and H^+ poles of the water molecules. Because of this tendency to break down into ions when placed in water, and ability to conduct an electrical current, sodium chloride is called an electrolyte.

Whenever a solid substance (such as NaCl crystals) breaks down in water solvent to create a saltwater solution, a higher degree of disorder results. The atoms (ions) tightly locked into a salt cube are extremely orderly. But when the cube dissolves and becomes part of a liquid (saltwater), the ions are much farther apart and move much more randomly. In general, whenever a solid substance dissolves into a liquid state, its atoms or ions will have a much more disorderly arrangement.

1, Disorder

The Body-Builders: Highly-Ordered "Skeletons" of Carbon Atoms

We have already mentioned that carbon atoms can covalently bond together to create long carbon–carbon (C–C) chains. They can also create branching trees or closed ring structures, as shown in Figure 4.4.

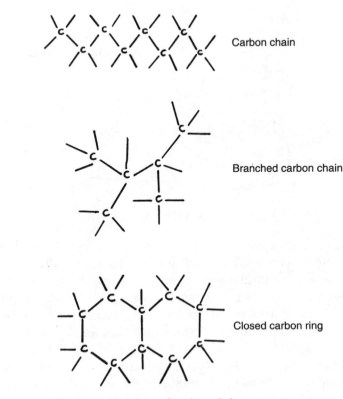

Carbon chain

Branched carbon chain

Closed carbon ring

Fig. 4.4 A variety of carbon skeletons.

Four Major Types of Chemical Body-Builders

The various groupings of covalently bonded carbon atoms form the skeleton or backbone of the four major types of chemical body-builders: *proteins*, *lipids* (**LIP**-ids), *carbohydrates* (**kar**-boh-**HIGH-draytes**), and *nucleic* (new-**KLEE**-ik) *acids*.

2, Order

PROTEINS

A protein is a large organic molecule consisting of linked *amino* (ah-**ME**-noh) *acids*. An amino acid is characterized by the presence of a nitrogen-containing amino (NH_2-) group of atoms. *Structural proteins* are the critical building-blocks found in most plant and animal tissues. *Enzymes* (**EN**-zihms) are proteins that speed up various chemical reactions, without themselves being changed in the process. Every living cell contains thousands of structural proteins and enzymes.

LIPIDS

A lipid is a group of "fats" (*lip*) and fat-like hydrocarbons that are not dissolvable in water. The main reason fats cannot dissolve in water is because they contain large numbers of uncharged C–C and C–H bonds, while water does not have bonds of this type. Many lipid molecules are created when too many calories are consumed. Various lipids also occur within the membranes of cells. Other members of the lipid family are really not fats, at all. This group includes the *hormones*, which act as chemical messengers.

CARBOHYDRATES

Carbohydrates are literally "carbon–water" molecules, that is, their molecules can be considered to consist of equal numbers of carbon atoms and H_2O molecules. Glucose, a very important carbohydrate, for instance, can have its *molecular formula* written as $C_6H_{12}O_6$, or as $C_6(H_2O)_6$ – six carbon atoms with six water molecules. In actual fact, however, glucose has a closed ring structure, with no water molecules included.

Some of the carbohydrates, such as glucose, serve as the major sugars used for energy by many cells. In certain plants, carbohydrates can help build cell walls.

NUCLEIC ACIDS

3, Order

The nucleic acids get their name from their occurrence within the "nucleus" of eukaryote cells. The nucleic acids include nitrogen-containing bases as their chief components. All of the living organisms contain two major types of nucleic acids – DNA and RNA. DNA is an abbreviation for *deoxyribonucleic* (de-**ahk**-see-righ-boh-noo-**KLEE**-ik) *acid*. *RNA*, on the other hand, is an abbreviation for *ribonucleic* (**righ**-boh-noo-**KLEE**-ik) *acid*. DNA occurs as a twisted ladder or double helix, and it includes genes along its length. These genes serve as codes for the production of body proteins and other chemicals. Several types of RNA make copies of the code, so it can be translated into the actual production of body proteins.

Chemicals for Metabolism

Energy is the ability to do work. All cells derive energy from their metabolism. They use this energy to carry out their required processes of life.

DIFFERENT TYPES OF ENERGY

2, Disorder

Technically speaking, several types of energy are usually identified. *Potential energy* is generally defined as energy that is locked-up or stored. This stored energy has the "potential" (possibility) to do work, providing that it is unlocked or released. Chemical bonds, for instance, contain potential energy. When these bonds are broken (as by the action of an enzyme), their stored (potential) energy is released. This released energy is frequently called *free energy* or *kinetic* (kih-**NET**-ik) *energy*, because it is "free" to help particles "move" (*kinet*), thereby doing some kind of work. One type of work done within cells is synthesis, making new molecules or organelles.

FREE ENERGY, WORK, AND THE ATP–ADP CYCLE

An important example of stored or potential energy is found within the bonds of the *ATP*, or *adenosine* (ah-**DEN**-oh-seen) *triphosphate* (try-**FAHS**-fate), molecule. The ATP (adenosine triphosphate) molecule is often symbolized as: **A-P~P~P**. The letter, A, of course, is an abbreviation for adenosine. Each letter P denotes *phosphate* (**FAHS**-fate), a phosphorus-containing chemical group. Within ATP, there are "three" (*tri-*) phosphate groups attached to the adenosine. The last two chemical bonds in the ATP are special high-energy bonds. This is indicated using the squiggly line, ~, before each of the last two phosphate groups (~**P**~**P**).

Many cells contain a special type of enzyme called *ATPase* (ATP-ace), literally meaning "ATP splitter" (-*ase*). ATPase enzyme, therefore, acts to split the second high-energy phosphate bond within the ATP molecule. When this bond is split, large amounts of previously stored potential energy is converted into free (kinetic) energy and is used to do work, such as synthesizing large proteins or other macromolecules. After losing the end phosphate group, ATP or A–P~P~P, becomes *ADP* or *adenosine diphosphate* (**die-FAHS**-fate): **A-P**~**P**. ADP, therefore, is a reduced version of ATP that contains "two" (*di-*) phosphate groups, rather than three.

When a person eats, say, a candybar or other foodstuff containing a high number of carbohydrate molecules (such as glucose), the individual carbohydrate molecules are eventually broken down. The potential energy stored in their chemical bonds is released. The free energy released from breakdown of food molecules is often used to re-attach the end phosphate group back onto ADP, thereby recreating more ATP. The food that most organisms either produce or eat, therefore, is eventually used to make more ATP. The cells of the organism then turn to the ATP, breaking it back down into ADP, such that more free energy is released to do the body's work.

The back-and-forth process between ATP and ADP can technically be called the *ATP–ADP cycle*. Whenever the cell is deficient in free energy, ATP is broken down into ADP. The released energy fills the gap and does cell work. Whenever the cell has an excess of free energy (such as after an organism eats a heavy meal), however, the opposite half of the cycle takes place. The excess free energy becomes stored as another high-energy phosphate group bond, converting ADP back into ATP. This ATP–ADP cycle is a continual process that goes around and around, for as long as the cell lives.

ANABOLISM VERSUS CATABOLISM: "BUILDING-UP" VERSUS "BREAKING DOWN"

The ATP–ADP cycle is an important part of the two primary processes of metabolism – *anabolism* (ah-**NAB**-oh-lizm) on one side of metabolism, and *catabolism* (kah-**TAB**-oh-lizm) on the opposite side. The distinction between these two opposite faces of metabolism is made clear by examining Figure 4.5. Anabolism is literally a "condition of" (-*ism*) "building up" (*anabol*), while catabolism, the exact opposite, is a "condition of casting [breaking] down."

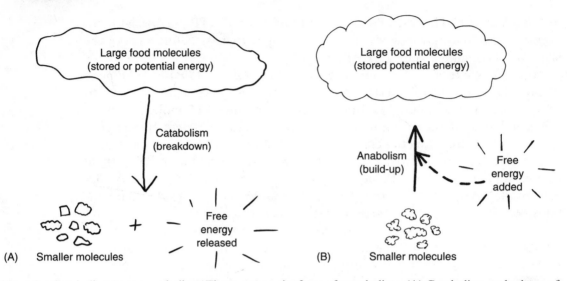

Fig. 4.5   Catabolism versus anabolism: The two opposite faces of metabolism. (A) Catabolism and release of free energy. (B) Anabolism and consumption of free energy.

3, Disorder

During catabolism (Figure 4.5, A), larger, more complex molecules of eaten food are broken down into simpler, smaller ones, usually resulting in the release of free (kinetic) energy. This free energy is then used to create ATP back out of ADP. And because more complex, larger substances are broken down into simpler, smaller ones, there is an increase in the amount of Biological Disorder present.

4, Order

During anabolism (Figure 4.5, B), the opposite process occurs. Smaller, simpler particles are brought together by consuming free energy released from ATP. New chemical bonds are formed, and larger, more complex molecules are synthesized. Such *anabolic* (an-ah-**BAHL**-ik) *processes* often result in the creation of new organelles or tissue structures, as well. Anabolism also plays an important role in tissue growth and repair functions. Due to the building up of larger, more complex body structures, anabolism is associated with an increase in the amount of Biological Order and pattern present.

CHLOROPHYLL AND PHOTOSYNTHESIS IN PLANTS

One critically important example of anabolism is the photosynthesis that occurs within plants. Plants are classified as *autotrophs* (**AW**-tuh-**trohfs**) – organisms that are "self" (*auto-*) "nourishing" (*troph*). Most plants are autotrophic or self-nourishing because they contain large numbers of *chloroplasts* (**KLOR**-uh-plasts). Chloroplasts are "pale green" (*chlor*) organelles "formed" (*plast*) within plant cells. Their green color is due, of course, to the presence of thousands of chlorophyll molecules. (Recall that *chlorophyll* literally means "green leaf.") The large amount of chlorophyll within leaves allows plants to produce most of their own free energy by themselves, using photosynthesis. In brief,

Cells of green leaves	$\longrightarrow$	Contain chloroplast organelles	$\longrightarrow$	Filled with chlorophyll molecules	$\longrightarrow$	Absorb light energy during photosynthesis

Photosynthesis or "synthesis" using "light" (photo-), involves two linked sets of chemical reactions. The first set is called the *light-dependent reactions*. The second set, however, does the actual synthesizing of chemicals. This second set is known as the *light-independent reactions* or *Calvin cycle*. In brief overview,

Photosynthesis = Light- + Calvin dependent cycle reactions (produces (produce energy) sugar)	*Study suggestion*: keep referring to Figure 4.6 as you read about the steps in photosynthesis.

In the first set of light-dependent reactions, the chlorophyll absorbs lots of energy striking the plant surface from sunlight. This energy absorption agitates the electrons within the chlorophyll. Some of these energized electrons are transferred, along with an H^+ ion from water, to a special *electron acceptor* molecule. Water (H_2O) molecules are split during this process. Thus, the molecules of oxygen (O_2) that remain after H^+ ion transfer are given off into the air. (This is the main reason why plants are oxygen-producing organisms.) The energy released by the transferred electrons helps add a phosphate group to ADP, creating more ATP.

Some of the produced ATP provides energy to run the second stage of photosynthesis. This second stage is called the Calvin cycle (after the scientist, Melvin Calvin). The Calvin cycle uses the ATP and special electron acceptor molecules that are given off by the first set of reactions.

The Calvin cycle is termed a "cycle" because it begins and ends at exactly the same point – carbon dioxide (CO_2) molecules. (This explains the well-known fact that plant cells consume CO_2.)

The cycle rotates once when it receives an input of one CO_2 molecule. ATP is also split to provide free energy. Several 3-carbon sugar molecules are produced. Some of these enter the *cytoplasm* (**SIGH**-toh-plazm) or fluid "matter" (*plasm*) of the plant "cell" (*cyt*). The 3-carbon sugars entering the cytoplasm are eventually converted into glucose, fats, or amino acids. The cycle has to rotate 6 times, and take in 6 molecules of CO_2, in order to synthesize one molecule of glucose. Finally, the glucose molecule serves as an important energy source for the plant cell.

The overall equation for photosynthesis is a simple one:

$$6\ CO_2 \quad + \quad 6\ H_2O \xrightarrow[\text{Light}]{} C_6H_{12}O_6 \quad + \quad 6\ O_2$$

Carbon dioxide + Water   Glucose + Oxygen

1, Web

Observe from the equation that photosynthesis releases large numbers of oxygen (O_2) molecules. Since ancient times, the green (chlorophyll-containing) organisms of this world have been steadily producing oxygen. Because of its large percentage of O_2 (about 1/5 or 21%), the Earth's atmosphere has become a true *biosphere* (**BUY**-oh-**sfeer**), or "ball of life." And practically all of the life in the biosphere (other than the green organisms) requires oxygen from photosynthesis for its ultimate survival.

GLYCOLYSIS AND RESPIRATION IN HETEROTROPHS

The preceding section discussed photosynthesis as an important example of anabolism or synthesis of glucose that occurred in the world of autotrophs, such as green plants. The glucose did not have to be eaten or ingested by the plants, since the plant cells and their chloroplasts manufactured it ultimately from the energy found in sunlight.

In *heterotrophs* (**HET**-er-oh-trohfs), however, the situation is markedly different. A heterotroph is an organism that is "nourished" (*troph*) by some "other" or "different" (*hetero-*) source. Human beings and most animals, for instance, are heterotrophs that eat or consume "other" organic foodstuffs (such as proteins, lipids, or carbohydrates) that come from "different" plants or living creatures. In short, when a human being needs a particular fuel molecule, such as glucose, it generally goes out and eats one as part of its diet! (The cells of heterotrophs such as humans can manufacture glucose or other fuel molecules by anabolism, but ultimately, they must still obtain some foodstuffs from other living, organic sources.)

The focus of this section, then, is not the anabolism or synthesis of glucose. Rather, the emphasis is upon *glycolysis* (gleye-**KAHL**-ih-sis) – the catabolism or "process of breaking down" (*lysis*) of "sweets" (*glyc*), such as glucose. Once glucose is present within the cytoplasm of heterotroph cells, its chemical

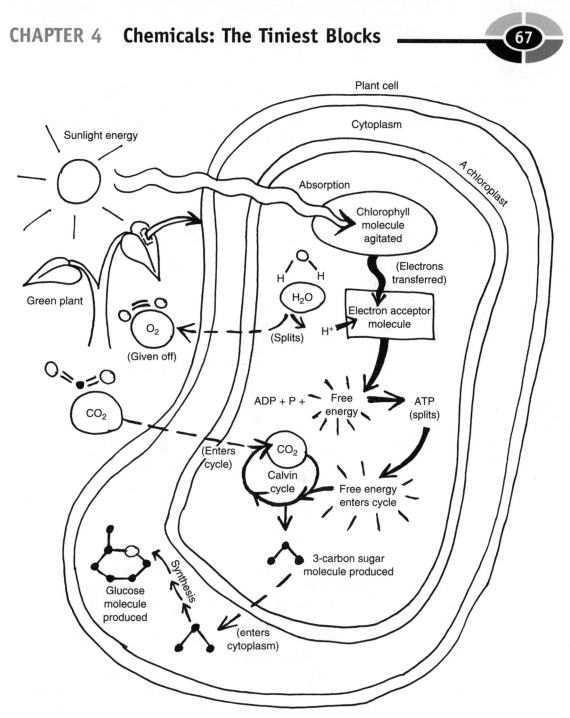

Fig. 4.6   An overview of photosynthesis.

bonds can readily be broken down by glycolysis with the help of enzymes, releasing lots of free energy to make ATP. During glycolysis, one 6-carbon glucose molecule is broken down into two 3-carbon *pyruvic* (pie-**ROO**-vik) *acid* molecules. Overall, as shown in the figure, 2 ATP molecules are produced.

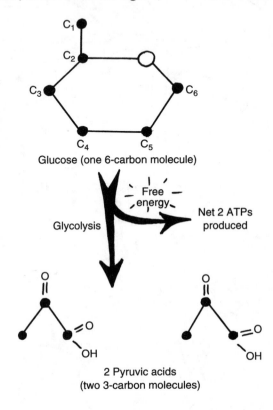

Glucose (one 6-carbon molecule)

Free energy

Net 2 ATPs produced

Glycolysis

2 Pyruvic acids
(two 3-carbon molecules)

In *anaerobic* (**an**-er-**OH**-bik) catabolism of a glucose molecule, this 6-carbon compound is broken down "without" (*an*) any oxygen from the "air" (*aer*) being present. During such conditions, the simple process of glucose decomposing into 2 pyruvic acid molecules (that is, glycolysis) is quickly followed by the conversion of the pyruvic acids into *lactic* (**LAK**-tik) *acid* molecules, as shown in the figure overleaf.

Lactic acid is named for its occurrence in sour "milk" (*lact*). But it is also produced by muscle and other body cells during anaerobic catabolism of glucose. When a human being runs very rapidly, for instance, the fibers (cells) in the leg muscles cannot get oxygen fast enough to allow them to utilize it for metabolism. Hence, glucose is broken down by anaerobic catabolism. Excess lactic acid molecules quickly accumulate, and they irritate local nerve endings within the muscle fibers. This creates leg cramping and pain.

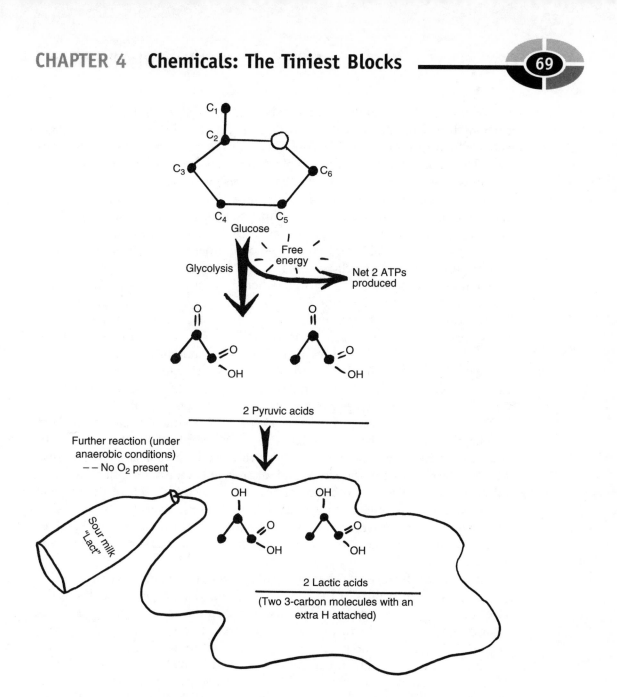

C_1
C_2
C_3
C_6
C_4
C_5
Glucose

Glycolysis

Free energy

Net 2 ATPs produced

2 Pyruvic acids

Further reaction (under anaerobic conditions)
– – No O_2 present

Sour milk "Lact"

OH

OH

2 Lactic acids
(Two 3-carbon molecules with an extra H attached)

When enough O_2 is present within the cell, however, glycolysis is followed by *respiration* (res-pir-**AY**-shun). Respiration literally translates from Latin as the "process of breathing" (*spir*), which is appropriate when it applies to the entire human or animal body. In the biology of individual heterotroph cells, however, respiration has a far different meaning. It is *cellular respiration*, not the process of breathing, itself, that is directly tied to cell metabolism. Cellular

4, Disorder

respiration is the breakdown or catabolism of organic molecules (such as carbohydrates, lipids, or proteins) within the cell, and in the presence of oxygen. Cellular respiration is therefore often called *aerobic* (air-**OH**-bik) *respiration*. This is because the process involves the breakdown of organic molecules within cells when oxygen from the "air" (*aero-*) is present.

Cellular respiration (aerobic respiration within cells) most often involves the catabolism of glucose or other simple sugars. Cellular respiration, then, starts up where the process of glycolysis stops. Glycolysis always provides the cell with a net of 2 ATP molecules/1 glucose molecule catabolized. Aerobic respiration provides the cell with many, many more.

To discuss respiration, we must go where it occurs – within the mitochondrion. You may recall (from Chapter 3) that the *mitochondria* (**my**-toe-**KAHN**-dree-uh) are the organelles where the cell's aerobic (oxygen-using) metabolism takes place. The mitochondria are literally "thread" (*mito-*) "granules" (*chondr*). This is evident from an examination of Figure 4.7. Some mitochondria are long and slender, like threads. Others are short and rounded, much like tiny granules. But whatever their shape, these mitochondria receive the two pyruvic acids produced from glycolysis.

Each pyruvic acid then enters the *Krebs cycle*. Named after the German biochemist, Hans Krebs, the Krebs cycle is a repeating cycle of aerobic reactions that break down the pyruvic acids produced by glycolysis. The enzymes that run the reactions of the Krebs cycle are located in the middle cavity of each mitochondrion. The Krebs cycle rotates twice (once for each of the pyruvic acids fed into it). Along the way, a net total of 2 ATPs, several CO_2 molecules, and a number of *hydrogen-carrier molecules* result. The hydrogen-carrier molecules, produced by the Krebs cycle then move onto the *electron transport system*.

The electron transport system is located along the *cristae* (**KRIS**-tee), the inner "crests" or "ridges" of the mitochondrion. The large molecules of the electron transport system, as their name indicates, carry high-energy electrons from H atoms down to lower and lower energy levels, releasing considerable amounts of energy along the way. A net total of 34 more ATPs are produced from the electron transport process. Finally, the transported electrons, now depleted of most of their former energy, are transferred to an oxygen atom. The oxygen atom then combines with two H^+ ions, thereby creating water (H_2O).

The overall equation for cellular (aerobic) respiration, using glucose as the fuel molecule, is:

$$C_6H_{12}O_6 \;+\; 6\,O_2 \;\longrightarrow\; 6\,CO_2 \;+\; 6\,H_2O \;+\; 36\;ATP$$

| Glucose | Oxygen | Carbon dioxide | Water | Free energy |

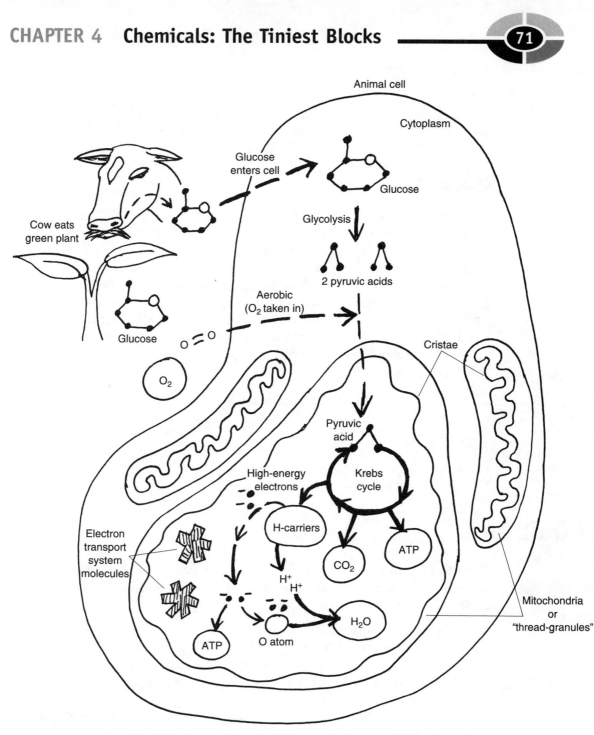

Fig. 4.7 An overview of cellular respiration (aerobic respiration).

Capsulizing this section, we have:

Glycolysis alone (anaerobic conditions) $\longrightarrow$ 2 ATPs/glucose
Glycolysis + cellular respiration (aerobic conditions) $\longrightarrow$ 38 ATPs/glucose

SUMMARY

Let us now put together what we have learned about metabolism in plants (autotrophs) versus animals (heterotrophs). Photosynthesis uses CO_2 and H_2O, but gives off O_2. Cellular respiration consumes the O_2 given off during photosynthesis, and gives off CO_2. These CO_2 molecules are then picked up by plant cells and consumed during photosynthesis. By this exchange of chemicals, photosynthesis feeds cellular respiration what it needs to proceed, while cellular respiration returns the favor. And so the Chemical Balance of Life between plants and animals, autotrophs and heterotrophs, goes on and on and on, generating and maintaining Earth's biosphere for hundreds of thousands of years.

2, Web

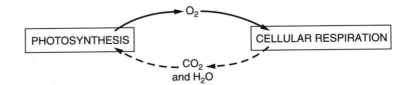

Quiz

Refer to the text in this chapter if necessary. A good score is at least 8 correct answers out of these 10 questions. The answers are listed in the back of this book.

1. Chemical bonds are:
    (a) Ways in which Biological Disorder is promoted
    (b) The means by which plants grow without need for any energy
    (c) Chemical linkages that serve as temporary storers of potential energy
    (d) Always due to a net loss of outermost electrons between different atoms

2. The four major types of chemical body-builders identified in this chapter are:

(a) Proteins, lipids, glucose, and benzene
(b) Lipids, fats, proteins, and carbohydrates
(c) Nucleic acids, inorganic salts, proteins, and lipids
(d) Carbohydrates, proteins, lipids, and nucleic acids

3. DNA and RNA molecules:
(a) Are nucleic acids found in all living organisms
(b) May help build proteins only in certain plants and fungi
(c) Are both located as sections of genes within cell nuclei
(d) Neither help nor hinder the synthesis of cell proteins

4. Kinetic energy:
(a) Represents free energy that is immediately available to do work
(b) Is basically the same thing as potential energy
(c) Is seldom useful for animals, just for green plants
(d) Is the form of energy locked up in the foods humans eat

5. Anabolism and catabolism differ in that:
(a) Anabolism is synthesis, whereas catabolism is breakdown
(b) Catabolism is synthesis, whereas anabolism is breakdown
(c) One involves only organic molecules; the other, only inorganic molecules
(d) Anabolism involves splitting of ATP, but catabolism never does

6. The overall equation for photosynthesis reveals that:
(a) Both oxygen and carbon dioxide are consumed, while H_2O is released
(b) Glucose and CO_2 are the ultimate products
(c) Sunlight provides the energy to produce carbon dioxide
(d) Carbon dioxide and water are the inputs, glucose and oxygen the outputs

7. The biosphere:
(a) Counterbalances the oxygen consumption of plants with carbon dioxide released from animals
(b) Represents the portion of planet Earth supporting living organisms within ecological systems
(c) Has no need for photosynthesis, as long as the sun is still shining
(d) Utilizes the blue-green algae as its sole free-energy source

8. In heterotrophs:
(a) Organic foodstuffs are consumed to obtain fuel molecules for ATP production
(b) Glycolysis is always the sole source of free energy for metabolism

(c) Energy from sunlight is utilized to agitate the electrons in chlorophyll

(d) ATP is split by ATPase enzyme, thereby consuming too much free energy

9. Under anaerobic conditions, 1 molecule of eaten glucose:
   (a) Ultimately results in 2 molecules of lactic acid
   (b) Immediately enters the Krebs cycle
   (c) Results in a total of 38 ATP molecules released for cell work
   (d) Provides free electrons for the cristae

10. These cellular reactions occur only within the mitochondria:
    (a) Photosynthesis by activation of chlorophyll
    (b) Both glycolysis and aerobic respiration
    (c) Aerobic respiration, Krebs cycle, and electron transport system
    (d) Calvin and Krebs cycles in heterotrophs

The Giraffe ORDER TABLE for Chapter 4
(Key Text Facts About Biological Order Within An Organism)

1. _____

2. _____

3. _____

4. _____

The Dead Giraffe DISORDER TABLE for Chapter 4
  (Key Text Facts About Biological Disorder Within An Organism)

1. _____
2. _____
3. _____
4. _____

The Spider Web ORDER TABLE for Chapter 4
  (Key Text Facts About Biological Order Beyond the Individual Organism)

1. _____
2. _____

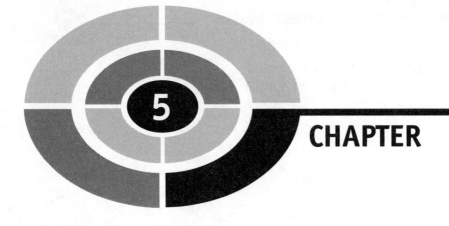

CHAPTER 5

Cells: The "Little Chambers" in Plants and Animals

We have already encountered cells (Chapter 2) as Level V within the Pyramid of Life. Figure 5.1 reminds us that this is the so-called "Life-line," meaning that the cell represents the lowest living level of biological organization. Review of this figure also reveals that Level I (subatomic particles), II (atoms), III (molecules), and IV (organelles), are included within the cellular level. *Cellular* literally "pertains to a cell or little cell."

The actual word, *cell*, in turn, comes from Latin and means "a chamber." This terminology arises from the early observation of dead cork under the microscope. Cork is actually the lightweight outer bark of the cork oak tree. When viewed through the microscope, dead cork is shown to consist of a complex latticework of hollow chambers with stiff walls. Each dark, hollow chamber in cork bark, then, is a type of non-living cell. But in general, however, the cell is defined as a living, tiny chamber that is surrounded by a thin membrane and contains various organelles.

1, Order

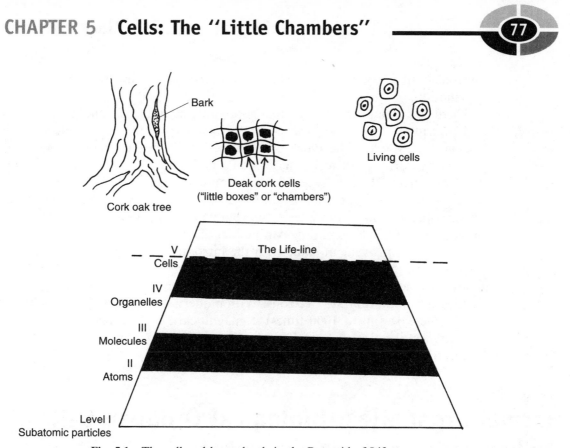

Bark

Living cells

Deak cork cells
("little boxes" or "chambers")

Cork oak tree

The Life-line

V
Cells

IV
Organelles

III
Molecules

II
Atoms

Level I
Subatomic particles

Fig. 5.1   The cell and lower levels in the Pyramid of Life.

Development of the Cell Theory

The topic of cells has been very important in biology since about the year 1665, with the studies of Robert Hooke, an English *microscopist* (my-**KRAHS**-cope-ist). Hooke used a simple microscope to view and draw the highly orderly pattern of rows and columns of hollow chambers within a thin slice of dead cork tissue. He described these hollow chambers as "little boxes" or "cells." But even though Hooke was the first person to use the term *cell*, for many years no one realized the true significance of these "little boxes" within living animal and plant tissues.

Over 100 years later, two German scientists, Schleiden & Schwann, took a giant step forward in human thinking. Based upon repeated drawings of animal and plant cells, about 1838 they advanced the *Modern Cell Theory*. This theory maintains that the cell is the basic unit of all living things. Therefore, to understand the structure and function of living organisms,

2, Order

one must ultimately study the cell level of biological organization for a logical explanation.

There was a major roadblock to a good understanding of cellular anatomy and physiology, however, for another 100 years after the Modern Cell Theory was stated. The chief problem was that not enough magnifying power and clarity of the cell interior could be obtained by using a *compound light microscope*. In this type of microscope, several or "compound" glass lenses help focus light and magnify viewed objects. With the compound microscope, not much more than the rounded cell nucleus ("kernel"), cytoplasm, and thin cell membrane can be identified.

A critical improvement came with the introduction of the *electron microscope* during the 1950s. Instead of just magnifying an object a few hundred times (like the compound light microscope did), the new electron microscope focused a beam of minute electrons. With modern instruments, this produces a huge increase (up to 1000 times) in magnification. As a result, organelles and large molecules such as DNA and proteins can now be directly seen within the cell.

Normal Organelles: Biological Order in Cells of Prokaryotes versus Eukaryotes

The modern electron microscope has been especially valuable in viewing the organelles of the prokaryote cell, such as a bacterial cell. Such cells are only about 1/10 the size of a typical eukaryote cell. Further, they are much simpler in their structural design. As you examine Figure 5.2, note that the prokaryote bacterial cell lacks a true nucleus. You may recall (Chapter 3) that the prokaryote cells appeared "before" (pro-) nuclei had evolved. Instead, the *bacterium* (bak-**TEER**-ee-um) holds a central, oval, *nucleoid* (**NEW**-klee-oyd) *region* that is "kernel-like," but not surrounded by its own individual membrane. The nucleoid region contains a complex collection of coiled DNA molecules. These DNA molecules use RNA molecules to help it direct and control the activities of the other organelles.

Prominent among these are the *ribosomes* (**RYE**-buh-**sohms**) or "5-carbon sugar" (*rib*) "bodies" (-*somes*). The ribosomes are tiny black bodies containing the 5-carbon sugar, *ribose* (**RYE**-bohs). These black bodies are the main locations for protein synthesis in the bacterial cell.

Like most other cells, the bacterial cell is surrounded by a soft *cell membrane* (also called *plasma membrane*). This cell (plasma) membrane encloses the cyto-

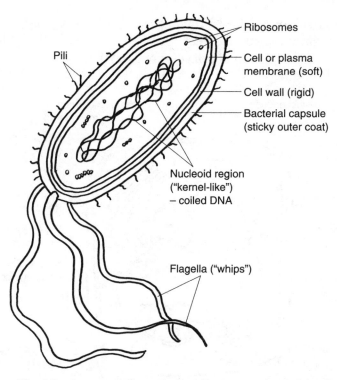

Pili

Ribosomes

Cell or plasma
membrane (soft)

Cell wall (rigid)

Bacterial capsule
(sticky outer coat)

Nucleoid region
("kernel-like")
– coiled DNA

Flagella ("whips")

Fig. 5.2   Anatomy of a typical prokaryote – a bacterial cell.

plasm and most other organelles. In addition, however, bacteria have a rigid
cell wall, a protective barrier outside the soft cell membrane. Most external of
all is the *bacterial capsule*, a sticky outermost coat that helps glue some types of
bacteria firmly to particular surfaces, such as those on human or animal cells.

Two types of projections often extend from the bacterial surface: *pili* (**PIE**-
lee) or short, "hair"-like strands, as well as *flagella* (flah-**JELL**-ah) or long,
"whip"-like strands. The pili help the bacterium attach itself to other objects,
while each *flagellum* (flah-**JELL**-um) acts like a whip to push the cell through
its watery surroundings.

EUKARYOTE CELLS: PLANT AND ANIMAL CELLS

Most plant and animal cells have a much more complex structure than does a
bacterium. As eukaryotes, both types of cells have their own well-defined
nucleus, surrounded by a *nuclear* (**NEW**-klee-ar) *membrane*.

Of the two eukaryote types, the plant cell has more in common with the
bacterial cell. Note from Figure 5.3, A, for instance, that the plant cell, like

the bacterium, is surrounded by a rigid cell wall. Likewise, plant cells have a plasma (cell) membrane immediately deep to the cell wall. This is quite different from the typical animal cell (Figure 5.3, B), which has only a plasma membrane surrounding it.

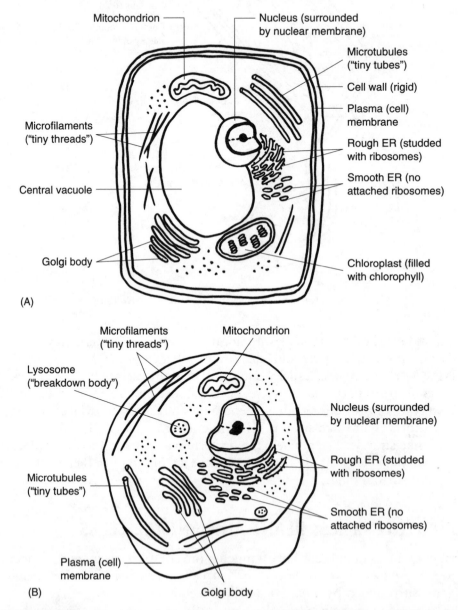

Fig. 5.3   Typical eukaryotes – plant and animal cells. (A) Plant cell anatomy. (B) Animal cell anatomy.

The plant cell has its green chloroplasts filled with chlorophyll, and a large *central vacuole* (**VAK**-yew-ohl). The central vacuole lies near the center of the cell, and it appears to be "empty" (*vacu*) or clear when viewed through a light microscope. In reality, however, it is a storage sac for various digestive enzymes. This makes the central vacuole in the plant cell the rough equivalent of the *lysosome* (**LIE**-soh-**zohm**) or "breakdown" (*lys*) "body" (*som*). Like the central vacuole, the lysosome contains digestive enzymes which, when released, digest or break down foodstuffs and other materials within the cell.

Both types of eukaryote cells have numerous mitochondria. These unique-looking organelles are frequently nicknamed the "powerhouse" of the cell, because they contain the chemicals necessary for aerobic respiration and ATP production. Also present in both cell types is an *endoplasmic* (**en**-doh-**PLAZ**-mik) *reticulum* (**reh-TIK**-yoo-lum). The endoplasmic reticulum (frequently abbreviated as *ER*), is literally a "tiny network present within the cytoplasm." The ER is a complex network of flattened sacs that serves to carry things around in the cell, much like a miniature circulation or highway system. The *rough ER* gets its name from the fact that its "rough" surface is studded with many ribosomes. The *smooth ER*, in marked contrast, does not have any ribosomes attached to its surface. The ribosomes on the rough ER engage in protein synthesis.

The synthesized proteins are often then circulated to the *Golgi* (**GOAL**-jee) *body or apparatus*. The Golgi body/apparatus is named after its discoverer, Camillo Golgi, an *histologist* (hiss-**TAHL**-oh-**jist**) or "one who specializes in the study of tissues." The Golgi body consists of a series of tightly stacked, flattened sacs. It mainly serves to package the proteins, lipids, hormones, and various other products of the cell.

Finally, both plant and animal cells contain a *cytoskeleton* (**sigh**-toh-**SKEL**-eh-ton). The cytoskeleton is literally the "skeleton" of the "cell," giving it some rigidity and support. The cytoskeleton consists of both hollow *microtubules* (**my**-kroh-**TWO**-byools) and solid *microfilaments* (**my**-kroh-**FILL**-ah-**ments**). The microtubules are "tiny tubes or tubules," while the microfilaments are just "tiny threads."

Protein Synthesis: Copying the Codons

Within the cell nucleus are a number of *chromosomes* (**KROH**-moh-sohms) or wormlike "colored" (*chrom*) "bodies." Each of these chromosomes, in turn, contains a number of tightly coiled DNA molecules. There are numerous genes or sections strung along the DNA molecules.

3, Order

A single gene provides a chemical code for the synthesis of a particular protein. As Figure 5.4 shows, part of the DNA double helix unwinds, exposing a group of *DNA codons* (**KOH**-dahns). These codons consist of sets of three chemical bases.

During *transcription*, a copy of the exposed DNA bases is made. A *messenger RNA* (*mRNA*) molecule then results. The mRNA molecule moves out of the nucleus, and onto the surface of a ribosome. A series of individual *transfer RNA* (*tRNA*) molecules, each attached to a certain amino acid, also move towards the ribosomes.

The final major stage of protein synthesis is called *translation*. During this process, which occurs along the ribosomes, the nitrogen base language of the

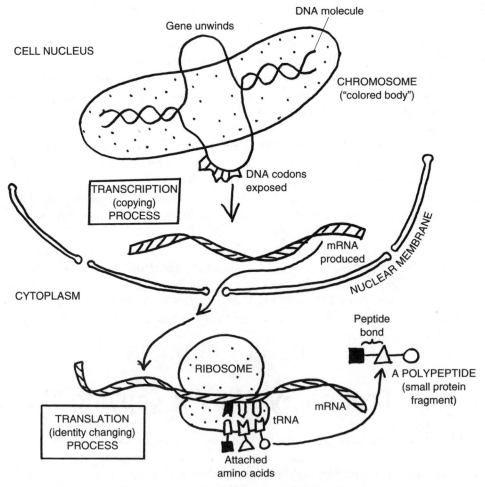

Fig. 5.4   Some basic steps in protein synthesis.

mRNA codons is translated or changed into the amino acid language of a certain protein.

The tRNA molecules match their bases up against complementary bases of the mRNA molecule. The amino acids attached at the other end of the tRNAs link together via *peptide* (**PEP**-tide) *bonds*. The end result is a finished protein or *polypeptide* (**PAH**-lee-**pep**-tide) – a combination of "many" (*poly-*) amino acids connected by peptide bonds in a coded order.

Each completed protein (polypeptide) detaches from a ribosome and begins to perform its special function within the cell.

Abnormal Organelles: Biological Disorder in Cells

Our earlier discussion of various normal organelles suggested the presence of Biological Order within cells. In general, such an order or pattern supports and maintains the health and survival of the cell.

What happens, however, when a cell organelle becomes abnormal or damaged? Certainly, one would expect to see some type of associated illness or cellular problem. Consider, for example, the condition called *cell autolysis* (aw-**TAH**-luh-sis) or "self breakdown" of a cell. When a particular cell is dying, dozens of its lysosomes may rupture simultaneously. This rupturing releases thousands of stored digestive enzyme molecules. When such a huge number of these powerfully dissolving enzymes are present at the same time, they break down the whole cell. Cell autolysis thereby often serves to remove dead or dying cells from otherwise healthy body tissue. Unnecessary (and perhaps disruptive) build-up of extra dead cells is prevented.

1, Disorder

Transport through the Cell Membrane

The cell organelles, whether normal or abnormal, require *transport systems* across the cell membrane to keep themselves functioning. Transport systems within the cell, like highway systems in a city, provide a nearly constant movement of particles or objects, some of them moving in, others moving out.

Transport systems are needed because the plasma membrane is a *selectively permeable* (**PER**-me-ah-bl) *membrane*. By selectively permeable, it is meant that certain types of particles are able to *permeate* (**PER**-me-ate) or "pass through" the cell membrane, while others are not able to pass.

There are two basic types of transport systems that move particles across the cell membrane: *passive transport systems* versus *active transport systems*. Passive transport systems are passive in the sense that they do not require any active input of ATP energy to function. Active transport systems, in direct opposite, are those systems in which free energy from splitting ATP energy is needed.

PASSIVE TRANSPORT SYSTEMS

There are three common types of passive transport systems serving cells. These are *simple diffusion*, *osmosis* (ahs-**MOH**-sis), and *facilitated* (fah-**SIH**-lih-**tay**-ted) *diffusion*. Diffusion is literally a "process of scattering" (*dif-fus*). The scattering process of diffusion arises from the fact that all particles are constantly moving in random directions. During simple diffusion, particles move by chance from a region where their concentration is high, to a region where their concentration is low.

Oxygen (O_2) molecules, for instance, tend to have a much higher concentration (crowding together) in the fluid outside of most cells, compared to their concentration within the cell. Why does this make sense? The reason is that oxygen molecules are being constantly used within cells for their aerobic metabolism. Thus, the *intracellular* (**IN**-trah-**sell**-yew-lar) *fluid* present "within" (*intra-*) the cell, generally has a low O_2 concentration. But the *extracellular* (**EKS**-trah-**sell**-yew-lar) *fluid* "outside" (*extra-*) the cell typically has a high O_2 concentration. Hence, there is a net (overall) simple diffusion of oxygen molecules from the extracellular fluid, across the plasma membrane, and into the intracellular fluid. [**Study suggestion:** Open a bottle of perfume. Place it onto a table in a quiet room. After half an hour or so, return to the room. Do you smell the perfume in the far corners of the room? Explain this observation.]

Closely related to simple diffusion is osmosis. Osmosis is a "condition of thrusting" (*osm-*). Specifically, osmosis is the simple diffusion of water (H_2O) molecules only, from a region where the water concentration is high, to a region where the water concentration is low. Think about a row of small green plants. In dry soil, their leaves and stems soon wither. But when the soil is freshly watered, the leaves and stems expand and stiffen. Osmosis of millions of H_2O molecules occurs from the moistened soil (having a high water concentration) into the cells of the green plants (which have a lower water concentration). The "thrusting" origin of the word osmosis therefore reflects *osmotic pressure*, the pushing or thrusting force associated with the simple diffusion of large numbers of water molecules.

Finally, facilitated diffusion (as its name strongly suggests) is diffusion that is facilitated or helped by the use of *protein carrier molecules*. Facilitated diffusion is basically just a process of "scattering" (like simple diffusion). It is special in that certain large molecules, such as glucose, need extra help in crossing the cell membrane after they "scatter" randomly to contact it. A *glucose carrier protein*, located right within the plasma membrane, combines with the glucose molecule. The carrier protein then changes its shape, dropping the glucose molecule off into the intracellular fluid. Here, the glucose can serve as an important source of cell energy.

ACTIVE TRANSPORT SYSTEMS

Active transport systems actively split ATP, unleashing free energy to power the transportation process.

Active transport, itself, is the ATP-requiring active pumping of particles from an area where their concentration is low, to an area where their concentration is high. Like facilitated diffusion, active transport uses a protein carrier molecule within the plasma membrane. Unlike facilitated diffusion, however, active transport runs "uphill" (energetically speaking), in that particles are moved from an area of low concentration "up" to an area of high concentration.

For example, sodium (Na^+) ions are removed from the intracellular fluid of many human and animal cells by an active transport system or "ATP pump." Sodium is at a much higher concentration in the extracellular fluid, compared to the intracellular fluid. Sodium ions rapidly pass through the plasma membrane and enter the intracellular fluid of nerve cells when a person is stimulated or excited. A sodium "ATP pump," consisting of a special carrier protein in the membrane, combines with the Na^+ ions that leaked into the cell, then splits ATP. This ATP-splitting generates enough free energy to actively carry Na^+ particles from the intracellular fluid of the nerve cells, back out into the extracellular fluid. A resting or nonstimulated state of the nerve cell membrane is thus re-established.

Cell Division and The Cell Cycle

Cells have a limited life span, and many of them are destroyed accidentally. Every single day, the human body loses millions of cells! Obviously, human beings, as well as most other multicellular organisms, greatly depend upon *cell division*. During cell division, one cell becomes split into two cells. The original cell is called the *parent cell*. The two cells resulting from its division are referred to as *daughter cells*.

THE CELL CYCLE AND MITOSIS

4, Order

Most cells within the human body go through the *Cell Cycle* – the entire life span of a particular cell, starting with its production from a previous parent cell, and ending with its division into two new daughter cells.

The Cell Cycle involves an orderly sequence of phases that is controlled by the DNA of the cell nucleus. *Interphase* is the phase occurring "between" (*inter-*) cell divisions. Interphase takes up the great majority of time (about 90%) in the Cell Cycle. "What is going on during this 90% of the time?" an inquiring reader would likely ponder. Interphase provides enough time for the cell to grow large enough to eventually divide into two living daughter cells. The cell also synthesizes numerous proteins, as well as additional organelles.

Interphase begins with *chromatin* (kroh-**MAT**-in), slender strands of DNA that have a dark "color" (*chromat*) and are covered with a "protein substance" (*-in*). The thin chromatin strands soon coil up and condense, creating thicker worm-like chromosomes. During interphase, the human cell makes copies of each of the 46 chromosomes in its nucleus. This creates 92 pairs of duplicated, identical chromosomes. These pairs are ready to be subdivided back into 46 single chromosomes, after the parent cell divides into two new daughter cells. The duplicated chromosome pairs line up in a vertical column in the middle or *equator* region of the parent cell.

After interphase, comes *mitosis* (my-**TOH**-sis). Mitosis literally means "a condition" (*-osis*) of "threads" (*mit*). The "threads," of course, are actually the thread-like chromosome pairs visible under a good microscope. The main "condition" that exists during mitosis is the division of the paired, duplicated chromosomes, into single, identical, unpaired chromosomes.

In preparation for cell division, a *mitotic* (my-**TAH**-tik) *spindle* is created near the nucleus. The mitotic spindle looks like an old-fashioned sewing spindle, being wider in the middle, and tapering towards both ends. (This really makes it resemble a modern fishing bobber!) The spindle is created from the orderly arrangement of cell proteins into a tapered, strand-like, pattern of microtubules. During mitosis, the duplicated chromosome pairs attach to the microtubules of the mitotic spindle. As the microtubules shorten, they pull the duplicated chromosomes apart from one another.

The separated 46 chromosomes are thus moved into opposite poles or ends of the cell. Finally, a *cleavage furrow* appears as a narrow groove between the two pinching-off poles of the parent cell. The cleavage furrow is eventually replaced on either side by a complete new cell membrane. The ultimate result – two separate daughter cells, each with an identical set of 46 single chromosomes of their own. Eventually, each of these daughter cells enters into its own Cell Cycle, beginning with another interphase. By this means, body growth and replacement of worn or damaged cells readily occurs within our human body tissues.

Cancer: A Severe Disorder of Mitosis

In a normal mitosis, the orderly duplication and division of chromosomes, along with the division of the cytoplasm, creates two identical daughter cells. Both of these daughter cells are normal, as well. Thus, they go about doing whatever body task they have been genetically programmed to perform. Epithelial cells, for example, frequently divide and replace themselves with daughters, each daughter cell in turn performing a body covering or cavity-lining function.

5, Order

But what can happen if an abnormal "change" (*mut*) or *mutation* (mew-**TAY**-shun) in the genetic program occurs? One of the really bad results can be *cancer*! The English meaning (and the astrological sign) for cancer, of course, is the "crab"! An alternate translation for cancer is "creeping ulcer."

Cancer is like a stubborn crab that can afflict either epithelial tissues (like the skin) or connective tissues (such as bone). When a person has a cancer, it is like a crab with pinchers, because it seems to hold on and not let go, making it very difficult to treat! Even worse, cancer has the characteristics of a creeping ulcer, because it often spreads from one affected site of the body, to many other sites. Small wonder, then, that various forms of cancer are a leading cause of death in many countries around the world.

2, Disorder

The exact cause of cancer in human beings, unfortunately, is still unknown. But various chemical agents, called *carcinogens* (car-**SIN**-oh-jens), are suspected cancer or "crab" (*carcin*) "producers" (*-gens*). Prominent among these suspected carcinogens are the poisonous or toxic chemicals present in cigarette smoke, which are thought to be the leading cause of lung cancer. The carcinogens may trigger abnormal changes (mutations) in the DNA of epithelial or connective tissue cells. The mutations create errors in the genetic program of the Cell Cycle, such that the resulting daughter cells are highly abnormal. The mutated, cancerous daughter cells

3, Disorder

have a largely unregulated, disorganized, and extremely rapid rate of mitosis. When these cancerous cells form larger tumors, the tumors interfere with the function and nutrition of normal cells. Eventually, the normal tissue cells die or are crowded out. If this process goes on long enough, the cell metabolism of the affected person may be so abnormally changed that the person dies.

Quiz

Refer to the text in this chapter if necessary. A good score is at least 8 correct answers out of these 10 questions. The answers are found in the back of this book.

1. The word, cell, comes from the Latin for:
    (a) "Crabs"
    (b) "Mice"
    (c) "Little people"
    (d) "A chamber"

2. The Modern Cell Theory:
    (a) Explains that the cell is the basic unit of all living things
    (b) Holds that, while cells are important, one must seek better explanations of anatomy and physiology at a different level of biological organization
    (c) Was not supported by drawings of tissues viewed through a microscope
    (d) Supports the concept that cell abnormality has no relationship to disease

3. An organelle always lacking from a prokaryote cell (such as a bacterium):
    (a) Ribosome
    (b) ER
    (c) Mitochondrion
    (d) Nucleus

4. Organelles that store digestive enzymes:
    (a) Lysosomes
    (b) Mitochondria
    (c) Nuclear membranes
    (d) Microtubules

5. ER:
    (a) Stands for "Emergency Reserves" within cells
    (b) Is a complex network of flattened sacs that circulates materials throughout the cell
    (c) Always has ribosomes attached to its surface
    (d) Has no known function within most cells

6. Codons:
    (a) Consist of particular chains of amino acids within proteins
    (b) Are not found within chromatin
    (c) Are chemical code words contained in genes
    (d) Essentially play the same roles as peptide bonds

7. During transcription:
    (a) Messenger RNA makes a copy of exposed DNA bases
    (b) Transfer RNA matches up with mRNA
    (c) Peptide bonds are created by linking nitrogen-containing bases together
    (d) A protein is manufactured from raw cellular materials

8. The cell membrane is described as selectively permeable, because:
    (a) All types of particles are able to freely enter and leave the cell
    (b) Certain particles are allowed to pass through it, while others are blocked
    (c) Millions of tiny ions constantly cross cell boundaries
    (d) No objects are allowed to permeate it

9. The overall movement of oxygen molecules from a high O_2 concentration in the extracellular fluid, to a lower O_2 concentration in the intracellular fluid is called:
    (a) Simple diffusion
    (b) Osmosis
    (c) Facilitated diffusion
    (d) Electrolyte imbalance

10. The longest portion of the Cell Cycle, during which protein synthesis and duplication of chromosomes occurs is known as:
    (a) Daughter cell cleavaging
    (b) Interphase
    (c) Mitosis
    (d) Normal cell stimulation by carcinogens

The Giraffe ORDER TABLE for Chapter 5
(Key Text Facts About Biological Order Within An Organism)

1. _____
2. _____
3. _____
4. _____
5. _____

The Dead Giraffe DISORDER TABLE for Chapter 5
(Key Text Facts About Biological Disorder Within An Organism)

1. _____
2. _____
3. _____

Test: Part 2

DO NOT REFER TO THE TEXT WHEN TAKING THIS TEST. A good score is at least 18 (out of 25 questions) correct. Answers are in the back of the book. It's best to have a friend check your score the first time, so you won't memorize the answers if you want to take the test again.

1. The water (H_2O) molecule:
    (a) Is an example of a crystal
    (b) Unequally shares outermost electrons between its O and H atoms
    (c) Frequently breaks down its peptide bonds to release energy
    (d) Equally shares outermost electrons between its O and H atoms
    (e) Is one of the least common types of molecules within most organisms

2. Both H_2O and $NaCl$ differ from CO_2 in that they:
    (a) Are important organic compounds
    (b) Do not contain chemical bonds
    (c) Represent key inorganic compounds
    (d) Never react with other particles
    (e) Are both mainly ionic in their bonding

3.   The main reason you shouldn't swim in salty water during a thunderstorm is:
     (a) Lightning might strike and topple a nearby tree, hitting you on the head!
     (b) Sodium chloride is very salty and irritating to the eyes and skin
     (c) Your body contains only organic solutes, while salty water contains only inorganic ones
     (d) Lightning might strike the water, and its electrons be carried to your body by ions
     (e) The rain might wash organic solvents into the surrounding seawater

4.   C–C bonds:
     (a) Can create long chains, branching trees, or closed-ring molecules
     (b) Always bond with hydrogen and oxygen at the same time
     (c) Are especially abundant in CO_2 molecules
     (d) Create a highly orderly inorganic skeleton for many body molecules
     (e) Seldom occur within the membranes of living cells

5.   General name for the fats and fat-like hydrocarbons that cannot dissolve in water:
     (a) Enzymes
     (b) Lipids
     (c) Carbohydrates
     (d) Krebs cycle electron carriers
     (e) Nucleic acids

6.   Type of energy released by action of ATPase:
     (a) Potential energy
     (b) Motivational juice
     (c) Kinetic energy
     (d) Consumption of calories
     (e) Cold stimulus

7.   Adenosine diphosphate:
     (a) Represents a reduced, lower-energy form of ATP
     (b) Continuously degrades into free energy, electrons, and O_2
     (c) Has no connection whatsoever to the Cell Cycle
     (d) Is regenerated whenever the cell has an excess of energy
     (e) May be important in humans and animals, but not in plants

8. You chew up and swallow a hamburger. Within your stomach and
    small intestine, it is further digested into individual proteins and amino
    acids, which in turn are finally broken down to provide free energy. The
    overall name for this process is:
    (a) Cannibalism
    (b) Anabolism
    (c) Tissue rehydration
    (d) Metabolism
    (e) Catabolism

9. After a portion of a tree trunk is gouged by a saw, the trunk slowly
    repairs itself, then continues growing. The general terms for these
    physiological operations are:
    (a) Mitosis & anabolic processes
    (b) Mitochondrial enzymatic action
    (c) Mitosis & catabolic processes
    (d) Stepwise decreases in Biological Order and pattern
    (e) Random osmosis/diffusion

10. Plant cells consume carbon dioxide as part of their metabolism because:
    (a) Cellular respiration always involves the net consumption of CO_2
    (b) The Calvin cycle starts with a group of C atoms coming from CO_2
    (c) Photosynthesis uses energy from sunlight
    (d) CO_2 molecules, not O_2 molecules, have enough energy in their
        bonds to make ATP
    (e) Most plants live in low-oxygen environments

11. All the greenish parts of a plant contain _____ with chlorophyll
    molecules:
    (a) Rough ER
    (b) Cell nuclei
    (c) Golgi apparatuses
    (d) Lysosomes
    (e) Chloroplasts

12. Cellular respiration:
    (a) Occurs right after glycolysis under anaerobic conditions
    (b) Utilizes O_2 to catabolize carbohydrates, lipids, or proteins
    (c) Does not occur in the cells of heterotrophs
    (d) Involves the production of lactic acid, under aerobic conditions
    (e) Takes place within the cytoplasm

13. The Krebs cycle:

(a) Produces hydrogen-carrier molecules, which then move onto the cristae of the mitochondria
(b) Is named for the concert pianist Krebs Kuhdiddlehopper
(c) Includes enzymes that operate directly upon the glucose molecule
(d) Is much less efficient than glycolysis in producing ATP molecules
(e) Has carbon as its final electron acceptor

14. Schleiden & Schwann are especially noted for:
(a) Advancing the Modern Cell Theory
(b) Being the first to use the word, "cell"
(c) Showing that cells were really not very important
(d) Painting pictures of organisms upon cave walls
(e) Successfully disproving the conclusions of Robert Hooke

15. The word lysosome translates into Common English to mean:
(a) "Breakdown body"
(b) "Kernel-resembler"
(c) "Lice-cutter"
(d) "Tiny digester"
(e) "5-carbon sugar body"

16. A packager of proteins, lipids, hormones, and various cell products:
(a) Lysosome
(b) Golgi body
(c) Nucleoid region
(d) Rigid cell wall
(e) Nuclear membrane

17. A cytoskeleton:
(a) Often cracks into tiny particles
(b) Provides pores in the nuclear membrane
(c) Manufactures antibodies
(d) Consists of microtubules as well as microfilaments
(e) Makes each cell extremely weak and fragile

18. A single gene is important because:
(a) It provides a code or blueprint for the production of a certain protein
(b) Most cells contain no more than one gene
(c) It always acts to speed up a particular digestive process
(d) Each gene provides a specific code for some type of membrane transport
(e) No DNA is present without it

19. If all the tRNAs within a cell were suddenly destroyed, _____ would be
     directly and immediately affected:
     (a) Osmosis
     (b) Transcription
     (c) Simple diffusion
     (d) Intracellular transport
     (e) Translation

20. A polypeptide gets its name from the fact that:
     (a) It consists of a series of unattached amino acids
     (b) The word mitosis also has the same meaning
     (c) The molecule really has a lot of "pep"!
     (d) The molecule contains peptide bonds between its amino acid
          subunits
     (e) There is nothing else quite like it within the living plant/animal
          body

21. One of the main reasons that body tissues do not become clogged with
     millions of dead cells:
     (a) Quick occurrence of autolysis
     (b) Microscopic "vacuum-cleaners" within the cytoplasm
     (c) Slow and reliable rates of structural protein synthesis
     (d) Just enough mitochondria to provide the needed energy
     (e) A massive allergic reaction to the dead cells

22. Active transport systems differ from passive transport systems by their:
     (a) Lack of importance in protein synthesis
     (b) Use of free energy from the splitting of ATP
     (c) Frequent lack of potential energy supplies
     (d) Stickiness and rubbery texture
     (e) Participation in cellular movements

23. Like simple diffusion, osmosis:
     (a) Proceeds from an area of higher to one of lower concentration
     (b) Occurs from an area of lower to one of higher concentration
     (c) Involves the random movement of H_2O molecules only
     (d) Directly releases energy to support cell metabolism
     (e) Interferes with normal cellular digestion

24. Facilitated diffusion shares a certain major feature with active
     transport:
     (a) Energy-free scattering of particles

(b) "Thrusting" of particles from a region where their concentration is low, up to a region where their concentration is high

(c) No similarity to simple diffusion

(d) Dependence upon the action of protein carrier molecules

(e) Breaking down of glucose

25. Mitosis is:

(a) The orderly division of paired, duplicated chromosomes into single, unpaired chromosomes

(b) A complete destruction of 46 chromosomes

(c) A specific type of cellular transport process

(d) Removal of mitochondria from the cytoplasm

(e) The same thing as interphase

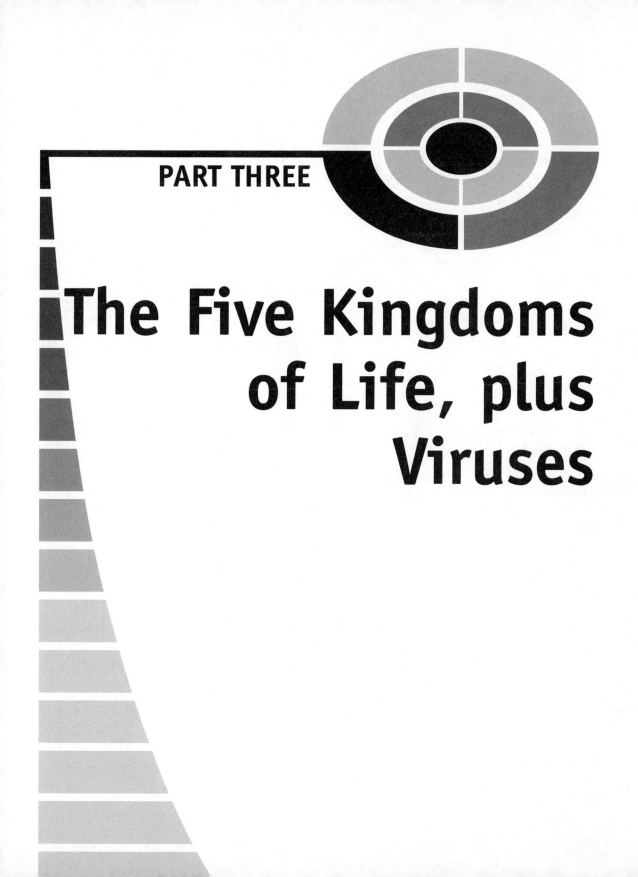

PART THREE

The Five Kingdoms
of Life, plus
Viruses

Bacteria and the "Homeless" Viruses

Bacteria as examples of prokaryotes (cells without nuclei) have already been introduced. Chapter 3 identified bluish-green bacteria as being among the first living things on Earth. And Chapter 4 further characterized the major organelles found within bacteria. It explained how these differed from the organelles found in the cells of most multicellular eukaryotes.

Chapter 6 now looks at the different types of bacteria in much more detail. This chapter also provides an overview of the major bacteria-caused disorders in infections of human, animal, and plant cells. Further, we will see how *viruses*, in turn, can infect both the cells of bacteria, as well as those of other living organisms.

Classifying and Ordering: The Work of Taxonomy

1, Web

Chapter 3 traced the evolution and appearance of major types of organisms in the Fossil Record. Besides ancient bacteria, there was discussion of plants, fungi, and animals. Such defining and classifying of different groups of organisms is the essential work of *taxonomy* (tacks-**ON**-oh-me). Taxonomy looks for general rules or "laws" (*nom*) for the "arrangement" or "ordering" (*tax*) of organisms into groups of various sizes. As such, it is an essential feature of Natural History (Chapter 1).

THE FIVE-KINGDOM SYSTEM OF CLASSIFICATION

2, Web

Ever since Aristotle and other early naturalists began observing and collecting observations about organisms, just how to classify them into particular groups for convenient study has been a major issue. Taxonomy involves no set laws or rules of Nature. Rather, it is a human-constructed way of assigning organisms to particular groups, using highly orderly and systematic methods.

Because it is human-made, there is no single rock-solid way of classifying organisms. Instead, there are a number of different ordering systems or kingdoms commonly used by *taxonomists* (tacks-**ON**-oh-mists). Perhaps the most simple of these methods is the **Five-Kingdom System**. As Figure 6.1 clearly illustrates, the Five-Kingdom System consists of three kingdoms of multicellular (many-celled) organisms, plus two kingdoms of unicellular (single-celled) creatures. According to this system, most multicellular creatures belong to either the *Kingdom Plantae* (**PLAN**-tie), *Kingdom Fungi*, or *Kingdom Animalia* (an-ih-**MAIL**-ee-uh). Obviously, these are the kingdoms of plants, fungi, and animals. And all single-celled organisms belong to either the *Kingdom Protista* (proh-**TIS**-tah) or *Kingdom Monera* (muh-**NIR**-uh).

Kingdom Protista consists of the *protists* (**PROH**-tists). Protists are often considered the "very first" (*protist*), that is, the most ancient, of all types of organisms. They are also among the simplest. The protists are eukaryotes, having a nucleus, but they are generally simpler than most plants, fungi, and animals. For example, the protist kingdom includes the *amoebas* (uh-**ME**-buhs). An amoeba is a single-celled eukaryote (Figure 6.1) that frequently "changes" (*amoeb*) the shape of its body as it moves through its environment.

Kingdom Monera is the one that includes the bacteria and all other types of prokaryotes. Since they lack a nucleus, all of the organisms within this

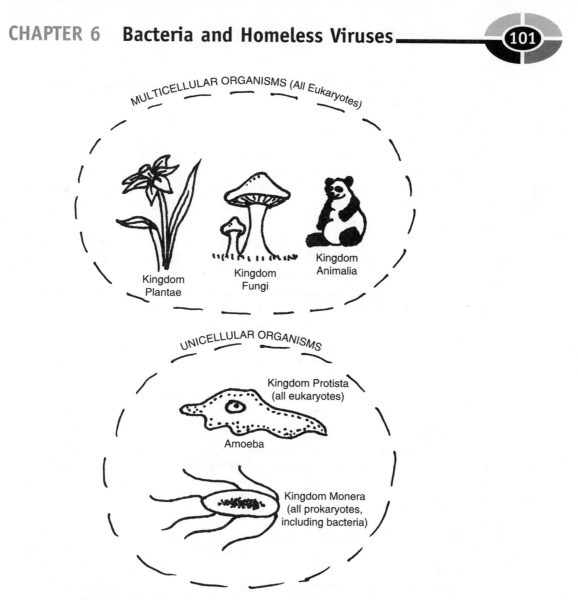

Fig. 6.1   The Five-Kingdom System of classification.

kingdom have cells with cytoplasm that essentially stands "alone" (*moner*). The bacteria and other *monerans* (muh-**NIR**-uns) also lack other membrane-enclosed organelles, such as mitochondria and Golgi bodies. Monerans thus have only the most basic facets of cell anatomy and physiology.

Despite small cell size and structural simplicity, however, the Kingdom Monera and its vast populations of prokaryotes (such as bacteria) make up the majority of the Earth's *biomass* or "living weight." All of the bacteria on this planet, added together, weigh far more than all the elephants and whales and human beings combined!

FINER CLASSIFICATION BELOW THE KINGDOM LEVEL

Taxonomists also recognize a number of levels of classification of organisms below the kingdom level. You may remember (from Chapter 2) that there is a Pyramid of Life. In this pyramid, the horizontal layers are the various levels of biological organization, starting with subatomic particles at the base, and finishing with an ecosystem at the peak. Taking a similar approach, Figure 6.2 shows a *Pyramid of Classification*. At the broad base of this new pyramid lies the *species*. A species consists of individual organisms of a certain "kind" (*species*). In a practical sense, two different organisms (male and female) are considered to be members of the same species if they can successfully reproduce to create fertile offspring.

3, Web

Above the species level lies the *genus* (**JEE**-nus). The word genus comes from the Latin for "stock" or "kind." A genus usually consists of two or more species belonging to the same "stock." This means that the related species making up a particular genus or stock have certain structural and functional characteristics in common. Further, these shared characteristics make the members of a particular genus distinctly different from any other group. In Kingdom Animalia, for example, we have the genus *Homo* (**HOH**-moh) or "man."

Taxonomists give a two-part Latin name to each species of organism. The first name (capitalized) is the genus, while the second is the species. We modern humans, for example, are classified as *Homo sapiens* (**SAY**-pee-enz) or "wise" (*sapiens*) "man." The human race, for all its problems, may not really be considered wise, but it is the only surviving species of the genus *Homo*. The Fossil Record has provided abundant evidence of other (now extinct) species within the *Homo* genus, such as *Homo habilis* (**HA**-bih-lis), *Homo erectus* (e-**REK**-tus), and *Homo australopithecus* (**aw**-stray-loh-**PITH**-eh-cuss).

Above the genus in the Pyramid of Classification lies the *Family* level. In taxonomy, a family consists of a group of related *genera* (**JEN**-er-ah, the plural of genus). Members of Homo sapiens, for instance, belong to the *hominid* (**HAHM**-ih-nid) or "man-shaped" family. Modern human beings and the man-like apes belong to the hominid family.

Next comes an *Order* of organisms. An order is a collection of related families or organisms. *Homo sapiens* belongs to the Primate Order, as do apes, monkeys, and lemurs.

Beyond the order is the *Class*. In taxonomy, a class is a particular group of related orders. Human beings and other members of the Primate Order, for instance, belong to the wider and more general *Class Mammalia* (mah-**MAY**-lee-ah) or "mammals."

Similar classes are grouped into a certain *phylum* (FIGH-lum). All Class Mammalia creatures are found within the still-larger category of the *Phylum chordata* (kor-**DAY**-tuh). The *chordates* (**KOR**-dates) are organisms with a slender "cord" in their backs, sometime during their development. All members of the Class Mammalia have such a slender cord, which eventually develops into a mature *vertebral* (ver-**TEE**-bral) *column*, or "jointed backbone," in the adult.

Finally, the highest taxonomic category of them all is the Kingdom. A kingdom consists of a group of related *phyla* (**FIGH**-lah, plural of phylum). All backboned organisms in the Phylum Chordata (including humans, of course), belong to the Kingdom Animalia.

4, Web

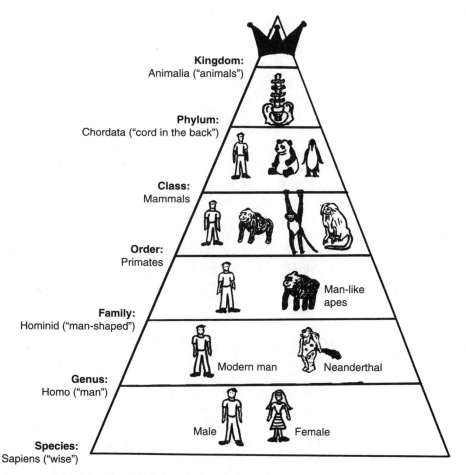

Fig. 6.2 The Pyramid of Classification for all types of organisms.

Classifying and Ordering Bacteria: The "Kings" of Monera

The Pyramid of Classification will be mentioned again and again, as we encounter various organisms in later chapters. In this chapter, however, the main emphasis will be on the characteristics of bacteria.

Most of the prokaryotes (monerans or members of the Kingdom Monera) known about, today, are bacteria. Hence, we can nickname the bacteria as the "kings" of Monera! One relatively simple way of classifying bacteria is the way they react to particular biological dyes or stains, such as the iodine-rich *Gram stain*. Bacteria are classified as being *gram-positive* if they stain violet with the Gram stain. They are called *gram-negative* if they lose the violet and take the color of the red opposite stain.

Another way of classifying bacteria and other monerans is by the general way in which they get their energy supplies. Recall from Chapter 4 that autotrophs are self-nourishing organisms, often relying upon photosynthesis for their ATP. And, likewise, remember that heterotrophs are nourished from some other source beyond themselves, usually by eating organic food-stuffs. Bacteria, too, can be classified as either autotrophs or as heterotrophs.

Among the autotrophs, there are two additional classifications, according to the specific way in which the bacteria are self-nourishing. *Photoautotrophs* (**FOH**-to-**aw**-tuh-**trohfs**) contain chlorophyll and use light and photosynthesis for their energy. *Chemoautotrophs* (**KEM**-oh-**aw**-tuh-**trohfs**) use chemical reactions, such as *nitrogen-fixation*, to produce their food. Nitrogen-fixing bacteria take nitrogen gas (N_2) out of the atmosphere and get energy by converting it into ammonia (NH_3) or ammonium (NH_4^+). Nitrogen-containing organic compounds like these, once produced by chemoautotrophs, are then used by green plants for their own metabolism.

THE MAJOR TYPES OF BACTERIAL SHAPES

In addition to classifying bacteria by the way in which they obtain their energy, they can also be slotted according to their shape (see Figure 6.3).

A *coccus* (**COCK**-us), or in plural form *cocci* (**COCK**-see), is rounded and sphere-shaped, like a "berry." A single berry-shaped bacterium is technically called a *micrococcus* (**MY**-kroh-**cock**-us). *Diplococci* (**DIP**-low-**cock**-see), in contrast, are "double berries." The diplococci are a group of spherical bacteria that occur in pairs (hence, doubles). *Strept* (**STREPT**) means "twisted."

Hence, the *streptococci* (**STREP**-toe-**cock**-see) are a genus of berry-shaped bacteria found as long twisted chains. Finally, consider the *staphylococci* (**STAF**-ill-oh-**cock**-see). These rounded, berry-like bacteria assume the shape of a "bunch of grapes" (*staphyl*).

In contrast to the rounded coccus form are the *bacillus* (bah-**SIL**-us) or "rod" shape and the *spirillum* (spy-**RIL**-um) or "coil" shape. Similar to the coil shape taken by the *spirilla* (spy-**RIL**-uh) is the "coiled hair" arrangement assumed by the *spirochetes* (**SPY**-row-keets).

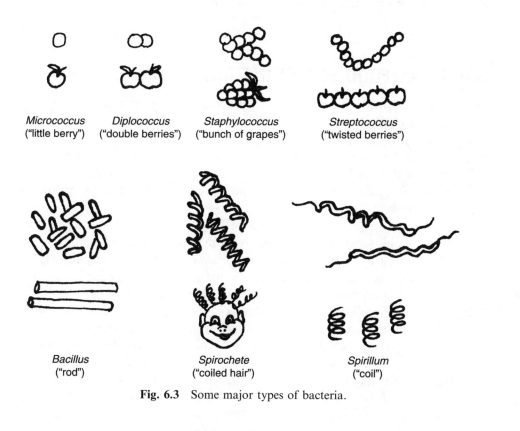

| Micrococcus ("little berry") | Diplococcus ("double berries") | Staphylococcus ("bunch of grapes") | Streptococcus ("twisted berries") |

| Bacillus ("rod") | Spirochete ("coiled hair") | Spirillum ("coil") |

Fig. 6.3   Some major types of bacteria.

Bacterial Order versus Disorder

Many types of bacterial species live in the soil, water, on the skin, and even within the intestines of human beings. Most bacteria participate in orderly relationships with their *hosts* (the creatures they live upon). There is often a condition of *symbiosis* (sim-be-**OH**-sis) – a successful "living together." For

5, Web

example, *Escherichia* (esh-er-**IKE**-ee-ah) *coli* (**KOH**-lie), often abbreviated as *E. coli*, is a type of bacillus commonly found in the *colon* (large intestine) of humans and other animals. Normally, *E. coli* lives in symbiosis with its human body host. These bacteria, along with many others, help produce certain B vitamins as well as sulfur-containing amino acids, which are then absorbed into the human bloodstream. In return, the *E. coli* benefit by consuming glucose and other organic molecules found within the large intestine. Hence, the individual human being successfully lives together with the millions of *E. coli* and other bacteria in the colon.

BACTERIAL FOOD POISONING AND BLOOD INFECTION

2, B-Web

1, B-Web

Many bacteria produce *exotoxins* (**EKS**-oh-**tahk**-sins) or "outside poisons," and then release these poisons into the surrounding environment. Even some types of *E. coli* bacteria, for example, when they are consumed in fecal-contaminated food or water, result in *bacterial food poisoning*. The accidentally consumed *E. coli* can release exotoxins, severely irritating the walls of the stomach and intestines, and giving the affected person a really bad case of *traveler's diarrhea*.

The other major problem associated with numerous bacteria are *endotoxins* (**EN**-doh-**tahk**-sins). These "inner poisons" are contained within the cell walls of certain gram-negative bacteria. *Salmonella* (**SAL**-moh-**nel**-AH) bacteria, for instance, are a genus of rod-shaped bacilli consisting of more than 1,400 separate species, most of which contain endotoxins in their cell walls. When such *Salmonella* are consumed in contaminated food (like raw chicken or hamburger), they may cause severe bacterial food poisoning.

3, B-Web

E. coli or other bacteria sometimes enter the bloodstream, perhaps through a dirty open wound in the surface of the skin. The result is *bacteremia* (bak-ter-**EE**-me-ah), alternately called *septicemia* (sep-tih-**SEE**-me-ah) – "a condition of bacteria or rotting" (*sept*) within the "blood" (*-emia*). Bacteremia (septicemia) in our normally sterile human bloodstream can create severe fever and widespread scarring and infections.

Finally, various health problems may result when the normal delicate balance among different types of bacteria within the human body is suddenly disrupted. Consider the excessive and prolonged use of general *antibiotics*, such as penicillin. Too much penicillin, taken for too long, can destroy beneficial bacteria and allow antibiotic-resistant strains of bacteria to develop. As these resistant bacteria multiply, they can create infections that are very difficult to treat.

Viruses: Non-living Parasites of Cells

In this chapter, we have introduced and briefly discussed the Five-Kingdom System commonly used to classify all *living* organisms. Note that in the preceding sentence, the word, living, was emphasized. The reason for this emphasis is the puzzling existence of a real organic oddball, the *viruses*.

The term *virus* comes from the Latin and exactly means "a poison." This name probably derives from the fact that a virus is a non-living superchemical that always invades living cells and, in a sense, "poisons" them by becoming a parasite. A virus, you see, cannot reproduce on its own. Therefore, it parasitizes human, animal, plant, or bacterial cells and uses their DNA/RNA to reproduce itself. In the process, the invaded cell is often destroyed, and the living organism becomes ill or dies.

1, Disorder

VIRAL STRUCTURE AND FUNCTION

A quick glance at Figure 6.4 (A) will quickly reveal why viruses are not considered living cells – they contain no plasma membrane or other organelles! A virus basically is a tiny parasitic particle whose simple structure consists of a core of nucleic acid surrounded by a coat of proteins. This extremely simple structure is enough, because viruses do not eat or drink, grow, synthesize proteins, or reproduce by themselves. Each viral particle contains either DNA or RNA as its nucleic acid, but not both of them. Recall that both DNA and several types of RNA are required for protein synthesis. Hence, viruses cannot make their own proteins.

1, Order

Helical (**HEE**-lih-kal) *viruses* contain nucleic acid wound up tightly into a coil or spiral, surrounded by a coat of small repeating proteins. *Polyhedral* (**PAHL**-ee-**he**-dral) *viruses* have a protein coat with "many" (*poly*-) triangular faces coming together. *Enveloped viruses* are enclosed by an outer lipid envelope. The strangest of the lot may be the *bacteriophage* (back-**tee**-ree-oh-**FAYJ**), which is sometimes just called a *phage* (**FAYJ**). Bacteriophage literally means "bacteria-eater"! While the bacteriophage doesn't exactly eat bacteria, it does attack and destroy many types of bacterial cells. The bacteriophage (phage) particle is topped by a multiple-faced *head* portion, a slender *neck* within a protein sheath, and several long *tail fibers* flairing out at the bottom. These tail fibers attach to the cell wall of the attacked bacterium, then inject viral DNA into it. (Examine Figure 6.4, B.) The injected viral DNA uses the bacterial host cell DNA and RNA to reproduce itself in huge numbers. Eventually, there may be so many new virus particles that they release enough powerful enzymes to cause the complete lysis (rupture and breakdown) of the infected bacterium.

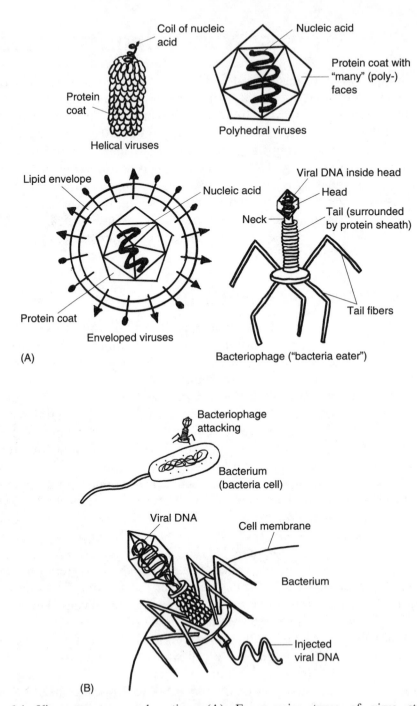

Fig. 6.4 Virus structure and action. (A) Four major types of virus structure. (B) Bacteriophage attacking a bacterium.

THE WORLDWIDE AIDS EPIDEMIC

Dangerous viral attacks upon human cells, not just bacterial cells, are unfortunately quite common. A prominent case in point is the deadly *AIDS virus*. AIDS is an abbreviation for *Acquired Immunodeficiency* (**im**-yew-no-deh-**FISH**-en-see) *Syndrome*. This disease is caused by infection of human cells with the *human immunodeficiency virus*, or *HIV*. The HIV particles are usually transmitted from an infected person to someone else during unprotected sexual intercourse. However, the virus may also be spread when the tears or saliva of an infected person go into another person's blood, or from blood-to-blood during a transfusion.

4, B-Web

The HIV particles are enveloped viruses that attack the *T-cells* in a human's *immune* (im-**YOON**) or "safety"-providing *system*. The viruses fuse with the T-cell's plasma membrane, then release their viral RNA into the cell. Eventually, the host T-cell produces DNA, which then directs the cell organelles to create new HIV particles. The new viruses bud-off from the surface of the host cell, and eventually infect many others.

Infection with HIV particles disturbs the immune or self-protective functions of the T-cell. When thousands of T-cells become infected, then, the immunodeficiency syndrome shows up. The AIDS virus, itself, does not kill the victim. Rather, it is the prolonged immunodeficiency of the infected person that is deadly. With very little *immunity* (protection) from disease in general, the AIDS patient becomes an easy victim for infection by many dangerous bacteria, such as those causing pneumonia.

Quiz

Refer to the text in this chapter if necessary. A good score is at least 8 correct answers out of these 10 questions. The answers are found in the back of this book.

1. Taxonomy essentially involves:
    (a) Study of surgical techniques for operating upon animals
    (b) Searching for general rules or laws for classifying organisms
    (c) Removing harmful species from particular ecosystems
    (d) Teaching people about the dangers or benefits of particular health practices

2. Often considered the very first (most ancient) and simplest of all groups
    of organisms:
    (a) Protists
    (b) Fungi
    (c) Plantae
    (d) Animalia

3. Kingdom _____ is the one including bacteria:
    (a) Amoebae
    (b) Protista
    (c) Monera
    (d) Wartae

4. Technical term for the "Family of Man":
    (a) Primates
    (b) Mammalia
    (c) Hominids
    (d) Vertebrates

5. Homo sapiens, apes, monkeys, and lemurs all belong to the _____
    Order:
    (a) Primate
    (b) Hominid
    (c) Homo
    (d) Mammalia

6. You scoop up a lump of wet muck and examine it through a
    microscope. Tiny amoebas slowly moving through the muck can be
    classified into what Kingdom?
    (a) Monera
    (b) Protista
    (c) Animalia
    (d) Fungi

7. The Phylum Chordata:
    (a) Includes all organisms with a slender cord or vertebral column in
        their backs
    (b) Does not involve any mammals
    (c) Chiefly encompasses the world of plants
    (d) Focuses upon the fungi

8. Berry-shaped bacteria arranged together like a bunch of grapes:
    (a) Streptococci
    (b) Diplococci

    (c) Bacilli
    (d) Staphylococci

9.   It is not wise to handle raw chicken, then lick your fingers, because:
    (a) Chicken skin is a deadly poisonous material
    (b) Septicemia is nearly inevitable
    (c) Too many foreign organic compounds may be ingested
    (d) Salmonella bacteria on the chicken may release endotoxins

10.   The basic structure of a virus:
    (a) Nucleus surrounded by a cell membrane and organelles
    (b) Core of nucleic acid surrounded by a protein coat
    (c) Just an envelope of DNA, nothing else
    (d) Fast-moving cell with hair-like projections

The Giraffe ORDER TABLE for Chapter 6
   (Key Text Facts About Biological Order Within An Organism)

1. _____

The Dead Giraffe DISORDER TABLE for Chapter 6
   (Key Text Facts About Biological Disorder Within An Organism)

1. _____

The Spider Web ORDER TABLE for Chapter 6
   (Key Text Facts About Biological Order Beyond The Individual Organism)

1. _____

2. _____

3. _____

4. _____

5. _____

The Broken Spider Web DISORDER TABLE for Chapter 6
   (Key Text Facts About Biological Disorder Beyond The Individual Organism)

1. _____

2. _____

3. _____

4. _____

7

The Protists: "First of All"

Now that bacteria (Kingdom Monera) and viruses (no kingdom) have been discussed, it is time to take a closer look at the protists. Figure 7.1 reviews the fact that unicellular organisms belong to one of either two major kingdoms – Kingdom Protista or Kingdom Monera.

1, Web

Protists = All Single-celled Eukaryotes

The protists include all of the single-celled organisms having a nucleus. This is a huge category, of course! And this category is distinguished from the bacteria, you may remember, which are the main single-celled organisms not having a nucleus. Recollect that Chapter 6 used the primitive amoeba as a representative member of the Kingdom Protista.

2, Web

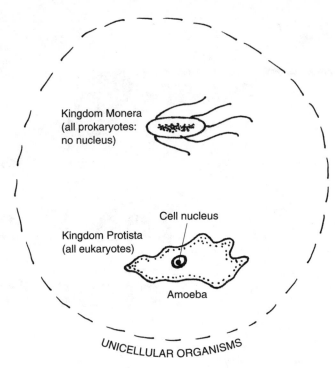

Fig. 7.1   The two kingdoms of unicellular organisms.

"WHERE DID THE FIRST CELL ORGANELLES COME FROM?"

Somewhere along the vast timescale of evolutionary history, a group of prokaryote cells without nuclei eventually developed into eukaryote cells containing nuclei. Recall (Chapter 3) that the first eukaryote cells appeared about 2.1 billion years ago. These original eukaryote cells started what we loosely call the protists, literally the "very first" organisms with nucleus-containing cells. Speculation about just how these nucleated protists first appeared has lead to the *Theory of Endosymbiosis* (**EN**-doh-**sim**-be-**OH**-sis). The concept of *symbiosis* (**sim**-be-**OH**-sis), in general, involves a "condition of" (-*osis*) "living" (*bi*) "together" (*sym*-). The term endosymbiosis, then, means a condition of living together and "within" (*endo*-). In a condition of endosymbiosis, two organisms of different species live together, with the smaller organism living inside the cells of the larger organism, which act as hosts. [**Study suggestion:** Picture a band of gypsies living together in a group of tents surrounded by a fence. The gypsies used to live independently, but for their safety and well-being, their band has come to live together with other members of a circus. The circus compound itself, including the fenced-off gypsy area, is surrounded by a tall wall. In return for food and protection, the gypsies live inside the circus compound and entertain their hosts (circus owners) and read their fortunes! The contained gypsy band and the larger host circus compound therefore live in a condition of endosymbiosis, where each group benefits.]

According to the Theory of Endosymbiosis, small prokaryote cells that used to live independently, moved into larger host prokaryote cells and achieved greater safety and well-being. (Examine Figure 7.2.) After living together in a state of endosymbiosis for a long time, the smaller prokaryote cells completely lost their ability to live independently. Instead, they evolved into nuclei and other organelles. Consider the possible origins of the mitochondrion, flagellum, and chloroplast. Mitochondria may have evolved from tiny, free-living, aerobic bacteria that were engulfed via phagocytosis by a larger anaerobic cell. Eventually, the bacteria mutated into mitochondria and

3, Web

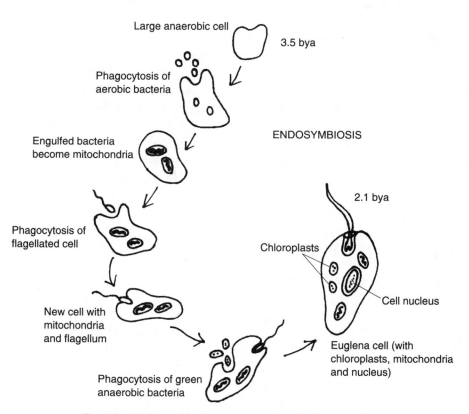

Fig. 7.2 Endosymbiosis and the origin of cell organelles.

became permanent residents and organelles of their larger host cell. By this means, the once anaerobic host cell became an aerobic (O_2-using) one containing mitochondria. Suppose that the now-aerobic cell then *phagocytosed* (**fag**-oh-**SIGH**-toesd) a small, fast-moving prokaryote with a whip-like flagellum. The entire small flagellated cell eventually evolved into a new organelle for the larger host cell – a flagellum. The host cell benefited by gaining increased mobility.

Similarly, the chloroplast and its capacity for photosynthesis may have evolved by endosymbiosis. Recall (Chapter 3) that tiny, bluish-green, chlorophyll-containing bacteria, shaped like threads or filaments, may have been the most ancient of living things. These bacteria were prokaryotes, lacking a nucleus. Many biologists speculate that early aerobic cells may have phagocytosed such bluish-green bacteria, which eventually evolved into slender chloroplast organelles. The large host cells would benefit by obtaining the capacity to produce energy via photosynthesis.

The ultimate result of all this endosymbiosis going on over long periods of time may well be such modern protists as *Euglena* (yew-**GLEN**-ah). This modern protist contains both mitochondria and chloroplasts as organelles, and it moves about rapidly by means of its flagellum. Commonly found in ponds, Euglena living in sunny water develop chloroplasts and become autotrophic, producing their own energy via photosynthesis. They use their flagella to scoot towards the light. In dark pond water, however, Euglena becomes an aerobic heterotroph, absorbing nutrients from the rich pond water and using its mitochondria to produce energy aerobically. Such dark-dwelling Euglena may even lose their chloroplasts!

So, is Euglena a plant (living anaerobically via photosynthesis), or is it an animal (living aerobically and using oxygen)? The answer is, "Neither!" For you see, Euglena is a protist!

DEBATE ABOUT THE KINGDOM

In recent years, there has been much debate among biologists about just whether the protists make up a particular kingdom, or whether the group should be split up into many separate kingdoms. The reason for this debate is the great differences among the 60,000 or so known species that make up the protist group. But there is one essential fact that all of the unicellular protists have in common. Although each of them contains a nucleus, the single cell of each protist is an entire organism! This makes the protists stand out distinctly from the nucleated cells of multicellular organisms, like humans, most plants, and animals. Human cells, for instance, typically have a nucleus, but they are

1, Order

also highly specialized members of particular body tissues, so they don't have to "do it all." Like the bacteria and other prokaryotes, however, the primitive protists were among the "very first" to appear on this planet, so that each cell must "do it all"!

The body structures and functions required for such cells to "do it all" will differ, of course, according to the habitat where they live, their main mode of nutrition, and the specific behaviors needed for their survival. The huge and varied group of protists, therefore, can be basically subdivided into three main subgroups. These are the *protozoa* (**pro**-toe-**ZOH**-ah), algae, and *slime molds*.

4, Web

The Protozoa or "First Animals"

Among the protists, it is the protozoa which most resemble tiny microscopic animals. Reflecting this fact, the word protozoa means "first animals." One of the main ways in which protozoa resemble animals is that they are heterotrophic. They must consume or "eat" other material as food. Typically, the protozoa ingest food (minute particles or other cells) via the process of phagocytosis.

THE RHIZOPODA OR AMOEBAS

One important phylum among the protozoa are the *Rhizopoda* (rye-**ZAHP**-oh-day) – technically, meaning the protozoa with "root" (*rhiz*) "feet" (*poda*). The amoebas are the major kind of *rhizopods* (**RYE**-zah-**pahds**). Amoebas get this name from their extensive use of *pseudopodia* (**SOO**-doh-**poh**-dee-ah) or "false" (*pseudo-*) "feet" (*pod*). As mentioned in Chapter 6, the word amoeba means "change." The amoeba seems to have its cytoplasm in a nearly constant state of change or streaming action, using its pseudopodia like blunt feet or root feet to scoot its body along and engage in phagocytosis of encountered food (see Figure 7.3).

The Algae: Plant-like Protists

The Fossil Record (Chapter 3) shows the first algae (literally, "seaweeds") appearing about 1.5 billion years ago. Algae are sometimes considered the most primitive members of the Plant Kingdom, whereas other scientists classify them as plant-like protists. Algae are fundamentally different from most

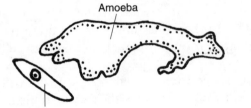

Amoeba

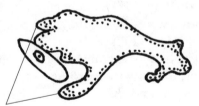

Another smaller
protozoan

Pseudopodia ("false feet")
surround protozoan

Beginning of phagocytosis
("cell eating")

Completion of phagocytosis

Fig. 7.3   The amoeba and its "false feet."

green plants, in that they lack any true leaves, roots, or stems. Primitive green algae are thought to be the forerunners of modern green plants. All algae contain chlorophyll, and therefore produce energy by photosynthesis. Most live in water, and many also have flagella to help them move. Some algae are tiny, unicellular, and microscopic, while others (such as certain giant seaweeds) exist as multicellular organisms many feet in length.

Unicellular algae include the *plankton* (**PLANK**-ton). They are free-floating algae that seem to be "wandering" (*plankton*) through the water of ponds or lakes or oceans. The plankton form huge floating masses that produce great amounts of oxygen. They also serve as an important food source for many *aquatic* (ah-**QUAT**-ic) or "water"-dwelling organisms.

An especially beautiful type of plankton are the *diatoms* (**DYE**-ah-tahms). The diatoms are single-celled microscopic algae with hard walls or *testas* (**TES**-tahs). These silicon-hardened testas must literally be "cut through" (*diatom*) to see the soft, tiny bodies housed within. Diatoms occur as about 10,000 different species, each species being covered by a unique and geometrically shaped testa. As evident from a quick scan of Figure 7.4, the diatoms show an extremely high degree of elegant geometric patterns and Biological Order at the microscopic level.

2, Order

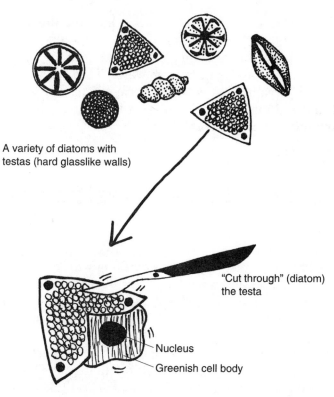

A variety of diatoms with
testas (hard glasslike walls)

"Cut through" (diatom)
the testa

Nucleus
Greenish cell body

Fig. 7.4   The beautiful geometric order of the diatoms.

While microscopic, unicellular plankton (including diatoms) float in the waters of the sea, another kind of algae, the *giant kelps*, are huge, multicellular brown seaweeds that can grow to over 150 feet in length! These mammoth brown algae often form extensive underwater kelp forests, providing both food and shelter for fish, sea otters, and other aquatic organisms. Surprisingly enough, however, the giant kelp do not have true stems, roots, or leaves. As multicellular algae, the giant kelp look somewhat like plants, but still have distinct differences from them. The kelp weed has leaf-like *blades*, which capture the energy in sunlight by conducting photosynthesis. Kelp is anchored to the ocean floor by root-like *holdfasts*.

The Slime Molds: Fungus-like Protists

The final major subgroup of the protists are the slime molds. Whereas the protozoa are considered the animal-like protists, and the algae are the plant-like protists, the slime molds are often nicknamed the fungus-like protists. The slime molds are thin, living masses of slimy material that use pseudopodia to move over damp soil, decaying leaves, and dead logs. Some of the 65 known species of slime molds live in fresh water. They receive their nutrients by phagocytosing organic matter from decaying plants. This feeding on decomposed material makes them similar to fungi, even though they move more like slimy groups of amoebas!

Biological Disorders: Some Parasitic Protozoa of Humans

Most of the billions of protist organisms dwelling with us on this planet are harmless, or even helpful, to humankind. Hard-working algae pump zillions of fresh O_2 molecules into our atmosphere every day, and even the lowly slime molds promote the decay of organic debris, so it does not accumulate to excessive levels.

1, B-Web

A relatively small number of protist species, however, are extremely dangerous to *Homo sapiens*, or introduce more Biological Disorder into the external environment. Consider, for example, *malaria* (mah-**LAY**-ree-ah) or "bad air." Malaria is a major disease of the tropics, and it kills over 2 million people every year! The disease gets its name from the humid, sometimes unhealthy, air found in tropical regions. But the real cause of the disorder has nothing to do with poor air quality! Rather, malaria is a deadly infection of human red blood cells (*RBCs*) with parasitic protozoa injected by the bite of a mosquito *carrier*. The infecting protists belong to the genus *Plasmodium* (plaz-**MOH**-dee-um), which means "mold" (*plasm*) "shaped." The Plasmodium organisms are actually parasitic protozoa that look somewhat like slime molds, in that part of their life cycle includes a stage that consists of a naked mass of cytoplasm containing many nuclei, but no cell membranes between them.

Part of the life cycle and infection of human RBCs with Plasmodium parasites is diagrammed in Figure 7.5:

- Step 1. An infected mosquito bites a human, injecting a number of *sporozoites* (**spor**-oh-**ZOH**-ites) – long, slender, tapered "seed animals" – into the victim's bloodstream with its saliva.
- Step 2. Within a few hours, the sporozoites travel through the bloodstream and invade cells in the liver. For 1 or 2 weeks, the sporozoites

keep growing and dividing and eventually develop into *schizonts* (**SKIZ**-ahnts), literally "dividing" (*schiz*) "beings" (*-onts*). It is this schizont stage which gives rise to the name Plasmodium, because each schizont is a fairly large mass of cytoplasm containing many nuclei.

- Step 3. Each large schizont then lives up to its name by splitting or dividing and releasing thousands of smaller parasite cells out of the liver, and into the bloodstream.
- Step 4. Each smaller parasite cell invades a human RBC, where it eventually grows into another large schizont.
- Step 5. The schizont filling the red blood cell eats the RBC from within, finally getting so large that it causes the RBC to rupture and be destroyed.
- Step 6. The rupture of thousands of RBCs releases toxins (poisons) and more small parasite cells into the bloodstream, causing high fever, chills, and other potentially fatal health problems.

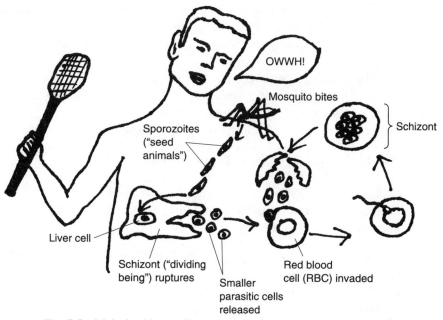

Fig. 7.5 Malaria: Plasmodium protozoa invade the liver and RBCs.

Quiz

Refer to the text in this chapter if necessary. A good score is at least 8 correct answers out of these 10 questions. The answers are found in the back of this book.

1. The two major kingdoms of unicellular organisms are:
   (a) Fungi and Protista
   (b) Animalia and Plantae
   (c) Protista and Monera
   (d) Mammalia and Herbivora

2. The Theory of Endosymbiosis implies that:
   (a) Gypsies were persecuted wrongly by the Nazis during World War II
   (b) Small, formerly independent prokaryote cells moved into larger host prokaryote cells and eventually evolved into organelles
   (c) Modern eukaryote cells could only have arisen from ancient bacterial cells already containing nuclei
   (d) Nucleated eukaryote cells lost their nuclei, becoming present-day bacteria

3. A junior high student on a field trip dips a tiny collecting bottle into the sunny water of a pond, and returns a sample to the science lab. Upon examining a drop of the pond water under a microscope, she will most likely see:
   (a) Rapidly moving, green-colored Euglena
   (b) Giant kelp and parasitic amoebas
   (c) Slime molds infecting aquatic bacteria
   (d) Brownish-colored Euglena without chloroplasts

4. This fact makes the unicellular protists stand out distinctly:
   (a) They all lack a nucleus
   (b) Each individual cell carries out the functions of an entire organism
   (c) Certain cells are specialized to perform particular limited functions
   (d) They cannot reproduce by themselves

5. Among the protists, it is the _____ which most resemble tiny animals:
   (a) Protozoa
   (b) Slime molds
   (c) Plantae
   (d) Algae

6. Pseudopodia:
    (a) May be utilized by both amoebas and slime molds for movement and feeding
    (b) Never appear in the anatomy of the protists
    (c) Literally translates as "root feet"
    (d) Help keep parasitic bloodsuckers from losing their grip on flesh!

7. Algae:
    (a) Are all unicellular and microscopic
    (b) Include giant kelp as well as diatoms
    (c) Closely resemble many amoebas
    (d) Can be dangerous parasites of human hosts

8. The main group of fungus-like protists:
    (a) Diatoms
    (b) Multicellular algae
    (c) Slime molds
    (d) Mold growing on stale bread

9. Protozoa are similar to most animals in that:
    (a) They are autotrophic in their feeding habits
    (b) Glucose and carbon dioxide are consumed during respiration
    (c) They are heterotrophic in their feeding processes
    (d) All types of organic and inorganic matter may be consumed

10. Symbiosis in general involves:
    (a) Cell division, resulting in tissue specialized for certain body functions
    (b) A condition wherein two different organisms live together, and they both benefit
    (c) One organism being a host to a parasite that causes it to become ill
    (d) Replacement of one organism by another more adapted to its environment

The Giraffe ORDER TABLE for Chapter 7
(Key Text Facts About Biological Order Within An Organism)

1. _____

2. _____

The Spider Web ORDER TABLE for Chapter 7
(Key Text Facts About Biological Order Beyond The Individual Organism)

1. _____

2. _____

3. _____

4. _____

The Broken Spider Web DISORDER TABLE for Chapter 7
(Key Text Facts About Biological Disorder Beyond The Individual Organism)

1. _____

The Fungi: Not Just Mushrooms!

Chapter 7 noted that the slime molds were fungus-like protists. Being thin, slimy, and amoeba-shaped, the slime molds feed on decomposing organic matter. This technically makes them *saprobes* (**SAH**-probes) or "rotten" (*sapr*) "livers" (*-obes*)! Many of the fungi, similar to the slime molds, are saprobes that release enzymes which partially digest the dead remains of previously living organisms. These digestive enzymes are largely responsible for the rotting or decomposition of dead organic matter. The saprobes then absorb nutrients as small molecules released by the digestion of dead, rotting remains. These saprobes are formally called the *saprophytic* (**sap**-roh-**FIT**-ik) *fungi*, since they get much of their nutrition from dead leaves and other parts of "plants" (*phyt*) that are "rotting."

The fungi were first introduced (Chapter 3) as plant-like organisms that are parasites of either dead or living organic matter. Thus, a difference between fungi and slime molds is that some fungi are parasites of living organisms. These *parasitic* (**pear**-ah-**SIT**-ik) *fungi* (as opposed to the saprophytic fungi) absorb small nutrient molecules from the cells of living hosts.

1, Disorder

Living human beings, for example, can suffer from *parasitic fungal* (**FUN**-gul) *infections* of their skin.

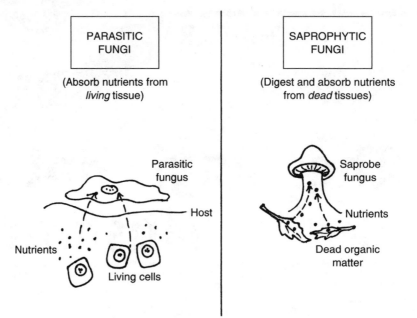

Many fungi are also pathogenic (disease-producing) in other living animals, as well as plants. Obviously, since they are parasites that live on others, fungi contain no chlorophyll and cannot carry out photosynthesis.

1, Web

The word fungus comes from the Latin for "mushroom." The Kingdom Fungi consists of molds and yeasts, as well as mushrooms (Figure 8.1). The yeasts are unicellular organisms, whereas both molds and mushrooms are multicellular ones. The general categories of molds, yeasts, and mushrooms all contain examples of both saprophytic and parasitic species of fungi.

The Basic Body Plan and Reproduction of the Fungi

In addition to the common characteristics already mentioned, there is also a basic body plan shared by most multicellular fungi. For illustration purposes, we will use a mushroom or *fruiting body* (Figure 8.2). The fruiting body of the mushroom releases many *spores* or "seeds" for reproduction. Millions of spores are given off from a mature mushroom or other fungus and float

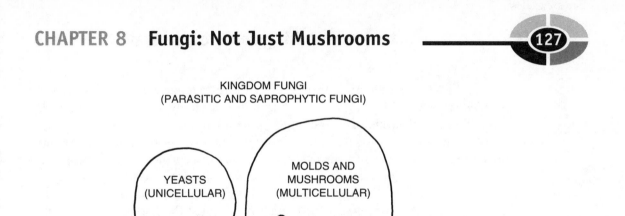

Fig. 8.1   The fungi: A kingdom of molds, yeasts, and mushrooms.

through the air, many of them falling down onto a neighboring area suitable as a food source. When the spore lands, it sprouts and starts absorbing nutrients. The spore grows out to create a *hypha* (**HIGH**-fah), a thin, thread-like filament. The word, hypha, comes from the Greek for "web." As its literal meaning suggests, each individual hypha branches repeatedly in a web-like manner as it grows. And two adjacent *hyphae* (**HIGH**-fee) of different *mating*

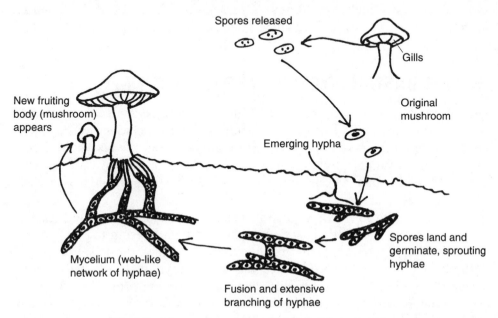

Fig. 8.2   The mushroom as a body plan for a fungus.

1, Order

types can fuse together and branch even more extensively. The result is a *mycelium* (my-**SEE**-lee-um) or "fungus nail," actually a web-like network consisting of thousands of branching hyphae or filaments. The mycelium of a mushroom is mostly located underground. The mycelium finally pushes up above the earth, creating the fleshy fruiting body or reproductive structure we call a mushroom. Quite surprisingly, what we consider a solid mushroom is actually an extensive web or network of hyphae (filaments), packed tightly together! And this umbrella-shaped mushroom is just the above-ground fruiting body (reproductive structure) of a much more broad and extensive mycelium hidden below the surface. [**Study suggestion:** Buy some fresh whole mushrooms and slice them thinly. Using a strong magnifying glass, look for a network of hyphae or slender filaments within each cap.]

The Major Groups of Fungi

More than 100,000 different species of fungi have been identified, yet *mycologists* (my-**KAHL**-uh-**jists**) – "those who specialize in the study of fungi" (*myc*) – usually group all of these species into just four different phyla. The names of these phyla are very tongue-twisting! Nevertheless, a helpful hint is that each phylum name ends with the suffix, *-mycetes* (my-**SEE**-teez), which comes from the Greek for "fungus" (*mycet*). So, look for and recognize -mycetes, and you know a fungus is somehow involved!

PHYLUM BASIDIOMYCETES (THE CLUB FUNGI)

A good fungal phylum to start with is the *Phylum Basidiomycetes* (bah-**sid**-e-oh-my-**SEE**-teez). The Phylum Basidiomycetes includes mushrooms, puffballs, and the *shelf fungi* often seen growing on the barks of trees. This particular fungal phylum, then, is quite familiar to those of us who like to walk through woods and fields (or even stroll through our own backyard).

The mushrooms get the first part of their phylum name from their *basidia* (bah-**SID**-e-uh) – the "little bases" (*basidi*) growing out from the gills under their caps (see Figure 8.3). This group is therefore called the Basidiomycetes ("little base fungi") or *club fungi* because of the tiny, "club"-shaped basidia present along the edges of their gills. The basidia are important because they produce the mushroom's spores, which can be blown off from under the cap and distributed long distances by the wind.

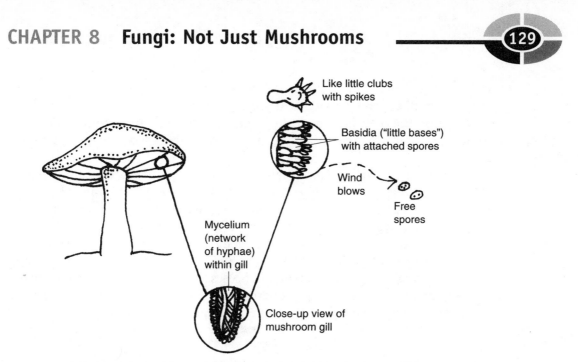

Fig. 8.3   The basidia: "Little bases" or "clubs" on mushroom gills.

PHYLUM ASCOMYCETES (THE SAC FUNGI)

You might also encounter various members of the *Phylum Ascomycetes* (**as-koh-my-SEE**-teez) or "leather bag" (*asco-*) "fungi" as you walk through a forest. This group is commonly called the *sac fungi*, because they produce spores within their sac-like or bag-like caps. Consider, for instance, the very popular *morel mushroom*, often nicknamed spongy-cap due to the wrinkled, sponge-like appearance of its cap.

Much less dramatic members of Ascomycetes, however, are the unicellular yeasts. Yeast cells reproduce by smaller cells budding off and eventually separating from larger ones. Yeast belonging to the genus *Saccharomyces* (**sak**-ah-roh-**MY**-seez) are the main "sugar" (*sacchar*) "fungi" (*myc*). This genus of yeasts has been used to help make bread, beer, and wine since ancient times. The yeast cells are added to raw bread dough or a vat of grape juice, and they soon begin to consume lots of sugar molecules in the dough or juice. Initially, these *Saccharomyces* organisms operate aerobically, utilizing oxygen. The yeast cells give off CO_2 gas as a waste product, and these gas bubbles are what make bread dough rise! Later, when the bread is finally baked, the yeast cells are killed by the high temperature of the oven.

During winemaking, in contrast, the sealed wine vats become anaerobic, after the active yeast cells consume all available O_2. The yeast cells then switch their metabolism from aerobic to anaerobic, and carry out the process

2, Disorder

of *alcoholic fermentation* (**fer**-men-**TAY**-shun). By this process, yeast cells produce both carbon dioxide and *ethyl alcohol* (common drinking alcohol) as waste products. When the fermenting wine reaches an alcohol content of about 12–16%, the yeast cells become poisoned and die from too much alcohol!

Some common types of molds are also considered members of the Phylum Ascomycetes (sac fungi). In general, a mold is a fuzzy coating of fungus growing on the surface of some food or animal or plant substances, when they are decaying or left for too long in a moist, warm place. A familiar example of such a fuzzy mold is the genus *Penicillium* (**pen**-ih-**SIL**-e-um), named from the Latin for "brush." The *Penicillium* fungi are bluish-colored molds growing on bread, fruits, and cheeses. They have a somewhat brush-like appearance under the light microscope, and several *Penicillium* species produce *penicillin*, the powerful, bacteria-killing, antibiotic drug.

PHYLUM ZYGOMYCETES (THE ZYGOTE FUNGI)

A third group of fungi are those of the *Phylum Zygomycetes* (**zeye**-go-my-**SEE**-teez), literally the "yoked together" (*zygo-*) "fungi." This phylum of fungi derives its name from the inclusion of a *zygote* (**ZEYE**-goat) within its life cycle. A zygote consists of two sex cells, called *gametes* (**GAM**-eats), that are literally fused, yoked, or "married" (*gamet*) together during fertilization.

The so-called *zygote fungi* in this group utilize *sexual reproduction* – the fusion of two gametes together to create a zygote. Once formed, the zygote then divides repeatedly by the process of mitosis (Chapter 5), thereby creating a new adult organism. The zygotes are enclosed in thick-walled *zygospores* (**ZEYE**-go-spoors), which are shed and sent wafting out through the air to new locations.

A typical *zygomycetic* (**zeye**-go-my-**SEE**-tick) fungus is the black bread mold, genus *Rhizopus* (**RYE**-zuh-pus) – "root" (*rhiz*) "feet" (*pus*). Interestingly enough, like many fungi, the *Rhizopus* group can engage in both sexual reproduction (gametes uniting to form zygotes), as well as *asexual* (**AY**-sex-you-al) *reproduction* that occurs "without" (*a-*) gametes. For simplicity, just the asexual reproduction of the black bread mold is displayed in Figure 8.4.

A spore lands on a piece of white bread, then begins to *germinate* (**JER**-muh-**nayt**). The landed spore "sprouts" a number of slender, thread-like hyphae, which soon merge to form a white, extensively branching mycelium, deep inside the bread slice. Soon, a large number of round-topped *sporangia*

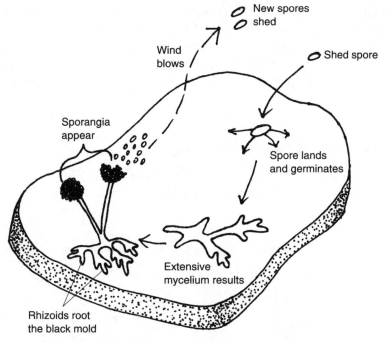

Fig. 8.4   The asexual life cycle of *Rhizopus* or black bread mold.

(spoh-**RAN**-jee-ah) or "seed" (*spor*) "vessels" (*angi*), appear like tiny black puffballs. The round, black sporangia are attached to the bread surface by long, narrow stalks. The mold is called Rhizopoda, because the sporangia and their stalks are firmly anchored into the bread surface by means of *rhizoids* (**RYE**-zoyds) – blunt, "root-resembling" or "foot-resembling" projections. When the wind blows, masses of black spores are scattered from the rounded sporangia, which hold thousands of them. A spore may land on another piece of bread, and the life cycle of the mold begins anew.

PHYLUM CHYTRIDIOMYCETES (WATER-DWELLING FUNGI)

A fourth phylum of fungi is the *Phylum Chytridiomycetes* (**KIH**-trid-e-oh-my-**SEE**-teez) or "water-dwelling" (*chytrid*) "fungi." The members of this group with an extremely tongue-twisting name are more easily called *chytrids* (**KIT**-rids), because they are mainly "water-dwellers."

Recent evidence suggests that the chytrids are the most primitive fungi, and that they were the first type to evolve from protists having flagella.

Dwelling in an aqueous (watery) environment, each chytrid has a small, globe-shaped body. It produces a highly active spore which has a flagellum attached. In their adult stage, the chytrids make a large mycelium (network) consisting of an extensive tangle of slender hyphae. These highly branched hyphae create a large area for the easy absorption of nutrients dissolved in the surrounding water.

Summary diagram of the fungi

Figure 8.5 provides summary pictures and brief descriptions of each of the four major phyla of fungi we have been studying.

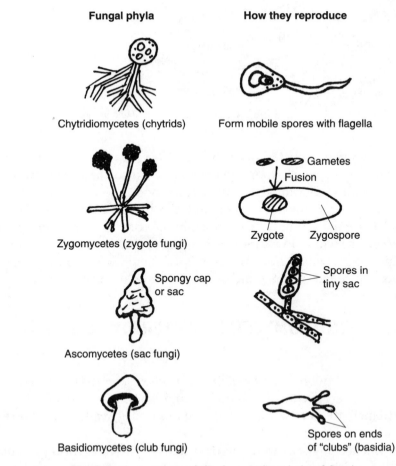

Fig. 8.5   A summary of the four major phyla of fungi.

The Lichens: Where Fungi Mix with Algae

Another unusual group closely related to the fungi are the *lichens* (**LIE**-kens). This term derives from ancient Greek and originally meant, "that which licks or eats around itself." Lichens are mixtures of fungi with either algae or bluish-green bacteria, living together in symbiosis (Chapter 7), such that each type of organism benefits.

2, Web

Because they contain photosynthetic algae that produce chlorophyll, lichens may look like greenish moss growing on a rock or tree trunk. Lichens are not green plants or really any other type of single organism. Lichens are constructed somewhat like a tough, stale, blueberry pie containing two different types of organisms (Figure 8.6). Like a pie, there is a protective upper and lower crust of slender, tightly packed hyphae (the fungus partner). As in the fruiting body (cap and stem) of a mushroom, the middle of the lichen consists of a mycelium – a loose network of highly branched, thread-like hyphae. Finally, like thousands of blueberries baked into a pie, small round algae are inserted into the meshwork of the fungus mycelium.

This relationship is *symbiotic* (**sim**-beye-**AHT**-ik) because the fungal crusts are tough enough to protect the contained algae from extremes of climate

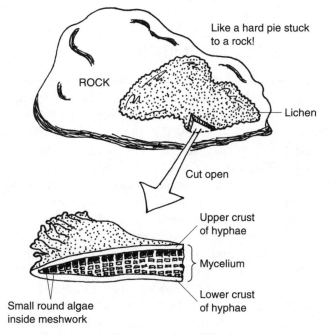

Fig. 8.6   The lichen as a slice of blueberry pie.

and temperature. The fungus also releases enzymes that partially digest rock or other hard surfaces, allowing nutrients to be absorbed by the lichen. In return, the algae produce additional energy for their fungal partners by carrying out photosynthesis. Thus, lichens make an extremely tough duo of cooperating organisms that, working together, can survive in some of the harshest habitats on planet Earth. Brave hikers can see them tightly clinging to rocks at the tops of frozen mountains, or even thriving in the empty wasteland of Antarctica!

Parasitic Fungi and Their Victims

Since certain fungi are parasites, it is only to be expected that some will cause the disease or death of their infected hosts. And those that are potentially pathogenic are nearly impossible to escape, because fungal spores float in the air, almost everywhere! [**Study suggestion:** Take the lid off an unwanted jar of jelly and leave it open. Soon, you will see the surface completely covered with a fuzzy substance. To what particular phylum in the Kingdom Fungi, do you think this fuzzy coating belongs? Its spores may have been produced and shed many miles away from you!]

FUNGAL PARASITES OF PLANTS

1, B-Web

Most of the parasitic fungi attack plants. *Dutch elm disease*, for instance, is due to a sac fungus (Phylum Ascomycetes) that was accidentally brought into the United States by infected logs from Europe during World War I. (The disease gets its Dutch name from the fact that it was first observed in Holland in 1919.) Spread from tree to tree by bark beetles, Dutch elm disease has eliminated thousands of native elm trees across the Continental U.S.

2, B-Web

Every year, grain farmers lose billions of dollars of crops to fungal infections. One grain infection with especially important implications for human health is *ergot* (**ER**-gaht), a small black sac fungus in the seed heads of wheat or rye plants. If any unwary humans or farm animals are so unfortunate as to eat such infected grain plants, they may come down with *ergotism* (**ER**-goh-tizm). During the Middle Ages in Europe, ergotism was quite common and was given the alternate name, *St. Anthony's fire*. People ate a diet heavy with rye and other natural grains baked in their bread. Although baking killed the fungus, it simply released its toxins into the bread. Such poisoning results in vomiting, cramps, great thirst, and burning in the abdomen. (Hence the old name, St. Anthony's fire.) Happily, modern grain processing techniques clean the ergot fungus from wheat and rye, so that ergotism is rare.

FUNGAL PARASITES OF HUMANS

Records show that when patients go to a hospital and accidentally acquire an infection, it is more likely to be a fungal infection than one of bacteria or viruses! And the offending parasite is most likely to be *Candida* (**KAN**-dih-dah) *albicans* (**AL**-bih-**kans**), a "glowing white" (*candid*) fungus. A close relative of the yeasts, *Candida albicans* is a common inhabitant of the human mouth, skin, intestine, and vagina.

3, Disorder

  Candidiasis (**kan**-dih-**DIE**-ah-sis), or an "infestation with" (-*iasis*) *Candida* fungus, tends to occur whenever the number of these fungi in the human body becomes very excessive. This may happen when a person is given a long-term regimen of antibiotics or steroid drugs, which depresses the activity of the *immune* (self-defense) *system*. Whenever too many competing organisms (such as bacteria) are killed off, or whenever the immune system becomes too weak, this leaves the door wide open for *Candida albicans* to overpopulate part of the body and cause a problem. Scattered white patches appear in infested areas, such as the female vagina, along with an inflammation. This is a common cause of *vaginitis* (vaj-in-**I**-tis) – "inflammation of the vagina" – in women of reproductive age.

Quiz

Refer to the text in this chapter if necessary. A good score is at least 8 correct answers out of these 10 questions. The answers are found in the back of this book.

1. Saprophytic fungi:
    (a) Absorb small molecules from the cells of living hosts
    (b) Are not common enough to be worth studying
    (c) Release enzymes that partially digest dead tissues
    (d) Are named from their habit of feeding upon the sap leaking from injured trees

2. A key difference between fungi and slime molds:
    (a) Some fungi are parasites of living organisms
    (b) Most slime molds act as parasites of living hosts

(c) Fungi obtain nutrients by digesting and absorbing organic matter

(d) There are far more species of slime molds, than of fungi

3. The entire group of unicellular fungi:
   (a) Molds
   (b) Yeasts
   (c) Morels
   (d) Lichens

4. The fruiting body of a mushroom:
   (a) Its extensive underground mycelium
   (b) Spores released into the air
   (c) A dense web of hyphae projecting above ground
   (d) The sweet, cool part that always tastes like an apple!

5. The fungal phylum to which morel mushrooms belong:
   (a) Zygomycetes
   (b) Basidiomycetes
   (c) Chytridiomycetes
   (d) Ascomycetes

6. The basidia are important to a mushroom, since they:
   (a) Directly absorb nutrient molecules from decaying organic matter
   (b) Produce and hold its spores
   (c) Branch extensively underground
   (d) Allow the fruiting body to be efficiently removed

7. Phylum of fungi involving the union of two gametes:
   (a) Ascomycetes
   (b) Chytridiomycetes
   (c) Zygomycetes
   (d) Basidiomycetes

8. Probably the most primitive group of fungi, and the first to evolve from protists:
   (a) The chytrids
   (b) Ascomycetes
   (c) Zooplankton
   (d) Bread yeasts

9. *Penicillium*:
   (a) Releases thick-walled zygospores
   (b) A bluish-colored genus of mold that produces powerful antibacterial agents

(c) Frequently appears on lawns as groups of fairy ring mushrooms
(d) Is a typical club fungus

10. A lichen:
(a) Is a type of green plant (roughly the same thing as moss)
(b) Exists as a symbiotic combination of fungi with algae or bacteria
(c) May be too delicate to live in harsh climates
(d) Basically is the fruiting body of an alga

The Giraffe ORDER TABLE for Chapter 8
(Key Text Facts About Biological Order Within An Organism)

1. _____

The Dead Giraffe DISORDER TABLE for Chapter 8
(Key Text Facts About Biological Disorder Within An Organism)

1. _____
2. _____
3. _____

The Spider Web ORDER TABLE for Chapter 8
   (Key Text Facts About Biological Order Beyond The Individual Organism)

1.   _____
2.   _____

The Broken Spider Web DISORDER TABLE for Chapter 8
   (Key Text Facts About Biological Disorder Beyond The Individual Organism)

1.   _____
2.   _____

The Plants: "Kings and Queens" of the World of Green

Chapter 7 considered the protists, including the algae – primitive, plant-like protists. And Chapter 8 took a look at the lichens as mixtures of fungi with either algae or bluish-green bacteria. Both algae alone, and in combination with fungi as lichens, have in common the presence of chlorophyll. This pigment, of course, is the main molecule responsible for photosynthesis, and for producing the greenish color of these organisms.

The Human Sole–Plant Connection

Recall (Chapter 3) that the word plant literally means "a sprout." It comes from the Ancient Latin term *planta*, which also translates to mean "sole of the foot"! Quite fascinating is the connection to another term *plantigrade* (**PLAN**-tuh-**grade**) – "walking" (*-grade*) on the "sole of the foot" (*planti*). Human beings, bears, and raccoons are all mammals that share the common trait of being plantigrade. Specifically, these mammals all walk with the entire sole of each foot firmly "planted" upon the ground! It seems likely that the early Greek *botanists* (**BAHT**-uh-**nists**), or "those who specialize in" (*-ists*) "plants" (*botan*), were the ones who made this connection. Thus, we have humans who walk in a plantigrade manner, sole of the feet placed flat upon the ground. And from this same ground, up sprout the green plants!

This close anatomic and naming connection between the sole of the human foot and the plants springing up from the ground has also been extended to the names of some individual plants. Consider, for instance, the plant genus called *plantain* (**PLAN**-tun). As Figure 9.1 shows, the *common or broad-leaved plantain* is a low-lying weed with large, light-green leaves growing from its base. Often invading lawns, the common plantain gets its name from the shape of its leaves, which are flat and broad like the sole of a human foot! One can say, then, that the flat leaves of the plantain have a very *plantar* (**PLAN**-ter) or "sole of the foot-like" appearance. These leaves can be crushed underfoot by a walking human, as well as eaten by them (blanched and sautéed in butter and garlic).

Sole of foot (plantar) Plantain leaf Plantigrade walking

Fig. 9.1 Feet, soles, and plantains: A human–plant connection.

Here Come the Land Plants!

Most botanists believe that modern land plants are the ancestors of primitive green algae that lived along the edges of lakes or oceans. Gradually (about 450 million years ago), the land plants emerged completely from the water, and spread across the Continents. As they evolved, they became more and more different from their primitive algae ancestors. Technically speaking, a plant is a multicellular organism that carries out photosynthesis, contains chlorophyll, and develops from an *embryo* (**EM**-bree-oh). An embryo is literally "a sweller." In general, then, an embryo represents the earliest stages of development of any organism. A *plant embryo* is a partial, undeveloped plant contained within a seed, that eventually "swells" into a full plant.

1, Order

There are more than 300,000 different species of plants in the world, and the vast majority are land plants (rather than algae). A few land-based species have apparently returned to the aquatic environment, including the slender *sea grasses*. But there are so many more species of land plants compared to algae that we have entitled this chapter *The Plants: "Kings and Queens" of the World of Green.*

THE TWO MAJOR GROUPS OF LAND PLANTS: CONTAINING "VESSELS," OR NOT?

Land plants have a huge amount of *diversity* (duh-**VER**-suh-tee) or "turning aside" (*divers*) from common traits. Nevertheless, a broad classification into two main groups of plants is possible. The great majority of species make up the *vascular* (**VAS**-kyoo-lar) *plants*, while a relatively few species comprise the *nonvascular* (**NAHN-vas**-kyoo-lar) *plants*.

1, Web

Vascular literally "pertains to" (*-ar*) "little vessels" (*vascul*). The vascular plants are also called the *tracheophytes* (**TRAY**-kee-oh-**fights**) – "plants" (*-phytes*) containing "rough arteries or windpipes" (*trache*). The vascular plants (tracheophytes) are the plants whose bodies contain many small hollow tubes that resemble blood vessels or windpipes (Figure 9.2, A). The hollow tubes create a vascular (vessel-rich) tissue, which runs up and down throughout the plant. The vascular tissue also branches outward to create dense patterns of *leaf veins* along the sides of the plant.

In direct contrast to the vascular plants or tracheophytes are the nonvascular plants. Their alternate name is the *bryophytes* (**BRY**-uh-**fights**), or "tree moss" (*bry-*) "plants" (*-phytes*). The nonvascular plants (bryophytes) have no internal vessels or leaf veins (Figure 9.2, B). They also lack true leaves, stems, or roots. This nonvascular type is called bryophytes (from Latin) because the mosses and other moss-like plants are the main examples.

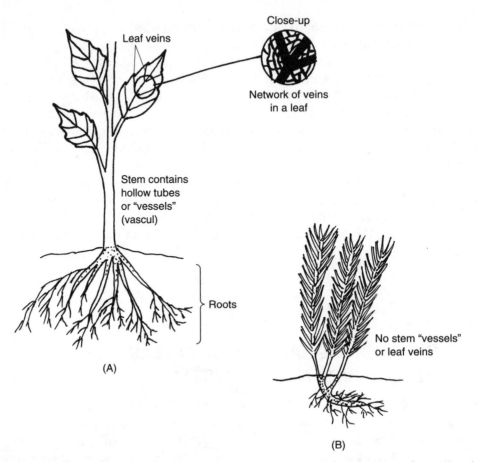

Fig. 9.2 Two major groups of plants: (A) Vascular plants (tracheophytes) – often tall and large. (B) Nonvascular plants (bryophytes) – short and small.

An Overview of the Bryophytes (Nonvascular Plants)

The two most familiar types of bryophytes are the mosses and the *liverworts* (**LIV**-er-werts). The word, moss, descends from the Old English for "bog" – a marsh or swamp consisting of soft, wet, spongy ground. A moss is a very small, soft, brown or green plant that grows closely together with others of its kind, forming a living carpet on rocks, trees, or ground. Mosses have short stems and many slender leaves. Decaying *sphagnum* (**SFAG**-num) *moss* (also called *peat moss*) is especially common in bogs. *Peat* (**PEET**) is heavy turf made of tightly packed mounds of partially rotted sphagnum (peat moss) and various other plants. Peat is used as a fertilizer and is commonly burned as fuel in Scotland, Ireland, and England, where there are many peat-containing bogs.

Liverworts are given two alternate names – *liverleaf* and *hepatica* (hih-**PAT**-uh-kuh). In humans and animals, the shorter word, *hepatic* (hih-**PAT**-ik), literally "pertains to" (-*ic*) the "liver" (*hepat*). Obviously, the words, liverwort and liverleaf, actually contain the smaller word, liver, while hepatica contains the Latin word for liver. A glance at Figure 9.3 will reveal the reason. The liverwort (liverleaf, hepatica) is a low-lying plant with brownish-green, three-lobed leaves that resemble the human liver, which also has a number of lobes. The resemblance of the hepatica plant leaves to the liver, created the ancient belief that eating this plant cures *jaundice* (**JAWN**-dis) and a number of other liver diseases!

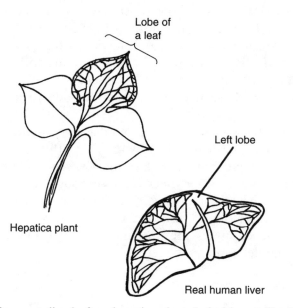

Fig. 9.3 The liverwort, liverleaf, or hepatica plant: Lobed leaves like the human liver.

LIVING HABITS AND REPRODUCTION OF THE BRYOPHYTES

Because both mosses and liverworts are bryophytes (nonvascular plants), they can live only in free-standing pools of water. Having no vascular (vessel-bearing) system or roots to transport fluid, nutrients, and waste products, they must grow upon pools of water (however tiny and shallow). Such a tiny pool might even occur in a shadowy cleft on the sun-baked face of a desert rock. A pool even this small may be just enough to allow the bryophyte plant to absorb water and nutrients into its nonvascular tissues, and survive.

A water pool is also needed as a fluid medium, so that sperm cells can travel from the male bryophyte plants and fertilize the female plants.

An Overview of the Tracheophytes (Vascular Plants)

2, Order

Figure 9.2 (A) showed some of the most basic anatomic features of the tracheophytes (vascular plants). In addition to their inner system of hollow passageways for transporting wastes and nutrients, this group of plants has true stems and roots. The two types of vascular tissue within tracheophytes are *xylem* (**ZEYE**-lem) and *phloem* (**FLOW**-em). Xylem comes from the Greek for "wood," because it often contains woody fibers that form the harder part of a plant vessel system. Xylem is stiff enough to provide support for vessel walls, and it also carries water and dissolved minerals upward into the plant after they have been absorbed by the roots. Phloem derives from the Greek for "bark." While the outer bark of most vascular plants is tough and dead, the phloem forms a softer inner bark composed of living tissue. The phloem (living inner bark) contains vessels which carry sugars and proteins synthesized in the leaves down towards the stem and roots.

Although all tracheophytes are vascular plants, some species produce seeds, while the simpler types lack seeds.

The Ferns: Vascular Plants without Seeds

Ferns are the main group of seedless vascular plants. A fern is a slender plant with feathery *fronds* (**FRAHNDS**) – split "leaves" that grow out along either side of a central stem. Ferns reproduce by means of spores. The spores grow in small brownish clusters on the back of the feathery frond leaflets. Ferns, huge fern-like trees, and their close relatives grew and reproduced by the billions in ancient tropical forests. Such forests of ferns covered vast stretches of the Earth during the *Carboniferous* (**car**-buh-**NIF**-er-us) *Period* of the

Paleozoic Era, 500 to 200 million years ago. Carboniferous literally "pertains to carbon-carrying." But it also has been nicknamed the "Age of Ferns." The Carboniferous Period gets its name from the great forests of ferns and related plants that grew during this time, died out, and were compressed into massive layers. After millions of years, the great pressure formed deep natural deposits of coal, oil, and natural gas. Because all of these "fossil fuels" are rich in carbon atoms, naming their time of origin the Carboniferous Period or "Age of Ferns" is very fitting.

2, Web

Gymnosperms: Vascular Plants with "Naked Seeds"

The great majority of vascular plants (other than ferns and their fern-like relatives) reproduce by means of seeds. One important group are the *gymnosperms* (**JIM**-nuh-**sperms**) or vascular plants with "naked" (*gymno-*) "seeds" (*sperms*). By naked, it is meant that the seeds of the gymnosperms are exposed, and not enclosed within fruits.

3, Web

As the Ancient World's climate became cooler and drier, many of the vast fern swamps of the Carboniferous Period died and dried up. Arising in their place were the *conifers* (**KAHN**-uh-**fers**) or "cone-bearers." The conifers, as their name suggests, produce cones that hold their naked seeds. Most of the conifers are evergreen trees or shrubs that bear cones. These conifers are truly "ever-green," because they keep their green, needle-like leaves all year. The thin leaves are an adaptation to cold, dry conditions in that their needle shape gives them a small surface area exposed to the air, thus reducing evaporation or drying of the conifer plant.

The conifers include pine trees, fir trees, spruce, larch, hemlock, and yew bushes. Today, much of the still-natural land of the Northern Hemisphere is covered by beautiful *coniferous* (kuh-**NIF**-er-us) forests, graced with tall spruce, fir, and pine trees.

THE LIFE CYCLE OF A PINE TREE

Since pine trees and other conifers are called gymnosperms, it is important to examine their typical life cycle, wherein their "naked seeds" play an important role. Each of the major steps are described below. (Consult Figure 9.4 for a summary.)

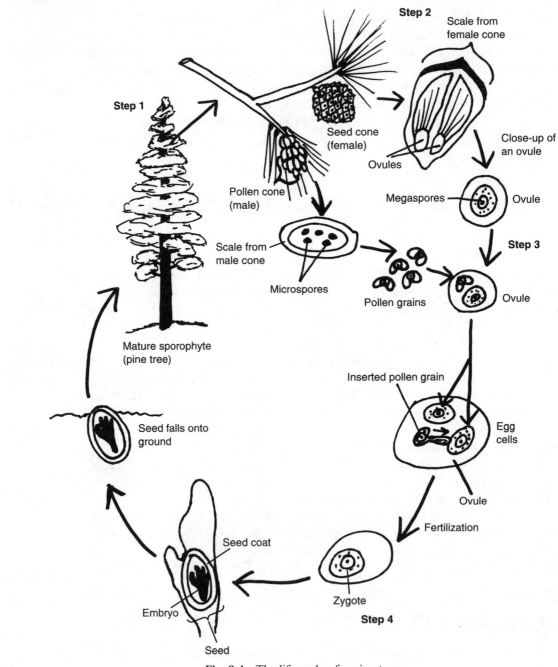

Fig. 9.4 The life cycle of a pine tree.

Step 1 A healthy mature sporophyte (pine tree)

An entire pine tree, itself, is technically called a *mature sporophyte* (**SPOR**-uh-**fight**). It is literally a "spore" (*sporo-*) or seed "plant" (*phyte*) that has matured. The same individual pine tree has two different sexes of cones hanging from its branches – small male cones, and not far away, big female cones. The small male cones are called *pollen* (**PAHL**-un) *cones*, while the much larger female cones are termed the *seed cones*. (It is the large female seed cones that we usually notice on the tree branches.)

Both types of cones produce spores for reproduction. Hence, when the male fertilizes the female, an interaction between their spores eventually results in a seed, which grows into a mature pine tree or sporophyte.

Step 2 Meiosis and spore production in both male and female cones

Pine cones produce spores by the process called *meiosis* (my-**OH**-sis). Meiosis means "a condition of lessening," wherein one cell divides into two daughter cells, each containing only half the number of chromosomes found within the original parent cell.

The small, male pollen cones consist of many loose, oval scales fastened together. Within each scale, meiosis produces tiny *microspores* (**MY**-kroh-spores). The microspores, each containing half the number of chromosomes of the parent tree, soon develop into gametes (sex cells). The male gametes become *pollen granules* – a "fine flour" (*pollen*) of tiny, yellow-colored grains.

The large, female seed cones consist of many tough, pear-shaped scales arranged into a spiral shape. The scales are woody and tilted outward at an angle. Each of these female scales holds two *ovules* (**OH**-vyools) or "little eggs." Each of these ovules (little eggs) produces a *megaspore* (**MEH**-gah-spore) – a "large" (*mega-*) spore.

3, Order

Step 3 Pollination (followed by fertilization) of female seed cones

When the male cones have matured, their scales pop open and release a cloud of pollen (millions of microscopic, yellowish grains). During *pollination* (**pahl-uh-NAY**-shun), a pollen grain lands on a female cone and enters one of its ovules. After pollination, meiosis is finally triggered within the ovule, and the megaspores eventually develop into female gametes, the egg cells.

Fertilization finally occurs when the pollen grain from the male grows out a tiny tube and releases a sperm cell into an egg cell.

Step 4 Zygote, embryo, and seed formation

The result of fertilization is a zygote (Chapter 8). The zygote undergoes repeated mitosis, adds many cells, and becomes an embryo. The original ovule within the female cone eventually develops into a pine seed, which contains the embryo. The seed provides continuing nourishment for the developing pine tree embryo, and it is covered by a tough seed coat. But the seed is still "naked" in the sense that it is not buried within a fruit. Finally, the seed falls to the ground, germinates (sprouts) under favorable conditions, and grows into a pine tree.

OTHER TYPES OF GYMNOSPERMS

Although conifers are the most common and widespread type of gymnosperms, there are several other varieties of seedless vascular plants. These are the *cycads* (**SIGH**-kads), *ginkgos* (**GING**-kohs), and *gnetophytes* (**KNEE**-toh-**fights**). A cycad is literally "a kind of palm." Cycads are not really palm trees, at all. Rather, they are a group of palm-like tropical plants with very long, fern-shaped leaves. About 100 species of cycads still survive. The group grew heavily during the Mesozoic Era (Chapter 3), when they shared the mostly warm, still-tropical planet with the dinosaurs. Cycad trees bear cones but, unlike pines, some individual trees hold only male cones, while other trees hold female cones only.

The word ginkgo comes from the Chinese for "silver apricot." The ginkgo genus is very ancient, but now there is only one surviving species. Like cycads, ginkgos are separated into either male trees or female trees. The male trees have cones, while the female ginkgo tree, as its name suggests, produces soft, fleshy seeds that look like apricots. However, these apricot-like seeds smell like rancid butter! The ginkgo tree has delicate, fan-shaped leaves that give it the nickname "maidenhair tree." Paleontologists and botanists have found fossils of ginkgos that are at least 200 million years old, yet are almost identical to the modern plant. Hence, the ginkgo (silver apricot, maidenhair tree) has sometimes been called a "living fossil." Ginkgo trees are not found in the wild. They have been grown domestically in China since antiquity, and now grace yards and streets in many cities worldwide as a popular ornamental tree.

Gnetophytes are a group of vines, shrubs, and small trees growing only in the tropics or in desert regions. Gnetophytes, like all gymnosperms, reproduce by means of spores and cones. One of the strangest gnetophytes is the *Welwitschia* (well-**WIT**-she-uh) plant. It is named after Dr. Friedrich Welwitsch (1806–1872), who first brought the odd specimen to Great Britain. The Welwitschia is only found in the deserts of Southwest Africa, where it appears as just two giant, strap-like leaves growing from the ground! Most of the huge, carrot-shaped stem is hidden below the ground surface, and it can grow to over 1 meter (3 feet) in diameter.

As a group, the gnetophytes are often treated as an evolutionary link between the conifers and the true flowering plants. The reason is that many types of gnetophytes bear cones (like conifers), but at the same time, the cones somewhat resemble flowers.

Angiosperms: Vascular Plants with "Flowers"

The gymnosperms, as we have seen, are vascular plants that reproduce by means of spores and seeds. The seeds are naked, and they are usually attached along the exposed edges of female cones.

Now we will examine the *angiosperms* (**AN**-jee-oh-**sperms**). These are the second major group of vascular plants, and they are named after an important part of their anatomy. These plants have "seed" (*sperm*) "vessels or holders"(*angi-*), meaning that their seeds are not naked. Rather, their seeds are enclosed within fruits. The angiosperms are also the true flowering plants, because their immature seeds are fertilized through a particular part of the flowers.

4, Web

The angiosperms probably first appeared on Earth during the Mesozoic Era, along with the gymnosperms, dinosaurs, and small mammals. [**Study suggestion:** Go back and review Table 3.1 in Chapter 3 to help you keep these evolutionary developments in perspective.] During these early days (245 million to 65 million years ago), the planet was carpeted with ferns (nonvascular plants) and conifers (seedless vascular plants).

As time wore on, however, the angiosperms (vascular plants with fruit-covered seeds and/or flowers) began to take over. How were they able to do this? The angiosperms (with flowers) had a unique reproductive advantage over the gymnosperms (without flowers). The advantage was the help of the insects, such as bees, moths, and butterflies, which go from plant to plant and assist with pollination. Further, certain animals (such as bats) and birds (for example, hummingbirds), also help pollinate flowering plants. Fruits, by

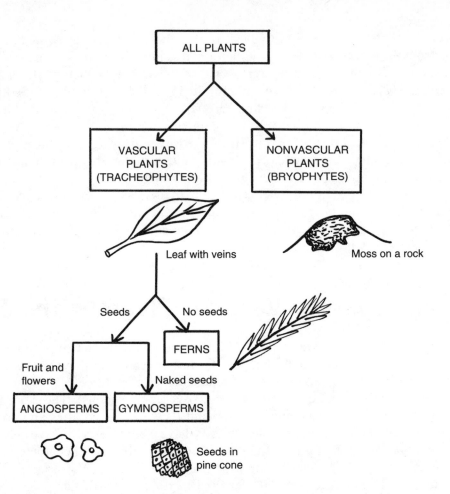

encasing their seeds, also assist the angiosperms in surviving and spreading. Now, within the Cenozoic Era, where we stand today (65 million years ago up to the present), the angiosperms have clearly become the dominant group of plants. There are about 275,000 species of flowering plants known, making them the most abundant and widespread, by far.

THE BASIC ANATOMY OF A FLOWERING PLANT

4, Order

Figure 9.5 reveals the basic structure of a flowering plant. Most critical, of course, is the anatomy of the flower, itself. A flower is the reproductive portion of a plant, the portion that produces the seeds. Anatomically speaking, a flower is a specialized body of modified leaves located at the end of a _pedicle_ (**PED**-uh-kul), a "little foot-like" stalk. The surface of this pedicle,

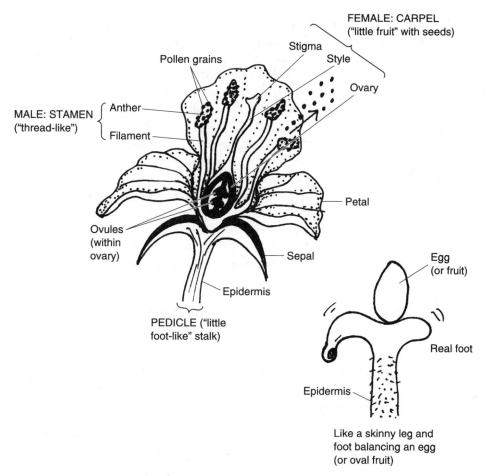

Fig. 9.5   The anatomy of a flower.

like the rest of the plant stem, leaves, and young roots, is covered by an *epidermis* (**ep**-ih-**DER**-mis). The word, epidermis, exactly translates to mean "(something) present upon the skin." In humans, the epidermis is the paper-thin outermost layer of the skin. In ferns and seedbearing plants, the epidermis is an extremely thin layer of outer protective cells.

In the center of the flower (which lacks an epidermis), lies the *ovary* (**OH**-vah-**ree**). It is the oval-shaped "egg" (*ov*) producer. [**Study suggestion:** Visualize a skinny human leg, covered by epidermis and sticking up into the air. At the top is the little foot, which has a deep downward curve on its sole. In the middle of the curve sits an egg. After you have visualized, check with the cartoon analogy and appropriate flower structures shown in Figure 9.5.]

Arranged around the bottom of the flower are two or more *sepals* (**SEE**-puls). These are green, leaf-shaped, outer "coverings" which often make a tight *calyx* (**KAY**-licks) or "flower cup" surrounding the closed flower bud, before it opens. Just above the calyx cup and its sepals are the *petals* – the thin, colored, plate-like portions of the flower that are "spread open." All of the petals together create a *corolla* (kuh-**ROLL**-ah). The corolla is literally the colored "crown" or "little wreath" (*coroll*) making a ring around the ovary. The petals often make a bright, colorful corolla that helps attract insects and hummingbirds to pollinate the plant.

Going inward, the next set of flower parts are the *stamens* (**STAY**-muns). The stamens are a group of thin, vertical, "thread-like" (*stamen*) filaments forming a tight inner ring that rises above the ovary. At the tips of the upright filaments, sit the *anthers* (**AN**-thers). The anthers are swollen, double-walled sacs that produce the male spores and pollen. The stamen, then, consists of slender filaments topped by anthers, and is the male reproductive organ of the flower.

STAMEN = FILAMENTS + ANTHERS = MALE REPRODUCTIVE ORGAN OF THE FLOWER

Finally, situated in the very center of the flower is the *carpel* (**KAR**-pull). The word, carpel, comes from Ancient Latin and means "little fruit." In botany, a *fruit* is the female part of the plant, which contains the seeds. The carpel (little fruit) consists of three parts: the *stigma* (**STIG**-mah), *style*, and ovary.

CARPEL = STIGMA + STYLE + OVARY = FEMALE REPRODUCTIVE ORGAN OF FLOWER ("LITTLE FRUIT" CONTAINING SEEDS)

The stigma and style form a sticky, green bulb that leads down into the ovary like a hollow pipe. The very center of reproduction – the ovule – is situated within the ovary.

REPRODUCTION OF A FLOWERING PLANT

Essential stages and processes in the life cycle of an angiosperm (vascular plant with fruit-covered seeds and flowers) are quite similar to those of the pine tree (a gymnosperm, vascular plant with naked seeds). The big difference, of course, lies with the fruit of angiosperms. The gymnosperms have naked seeds on the edges of their female cones. In angiosperms, however, the seeds are produced within its flowers, and then packaged within its fruits. A fruit, then, is simply a mature ovary of a flower, whose ovules or eggs have become seeds covered by an outer flesh. In some cases, the fruit is edible by

5, Order

humans and other animals (like a pea pod or an apple), while in other cases it is not edible.

For an example, consider the flower of a pea plant which, after pollination, gives rise to the pea pod and its peas (Figure 9.6). Everything depends upon pollination. If a flower has not been pollinated, then it will eventually wither and die, with no fruit being produced. Most flowering plants reproduce by *cross-pollination*. This means that the plant does not self-pollinate its own

5, Web

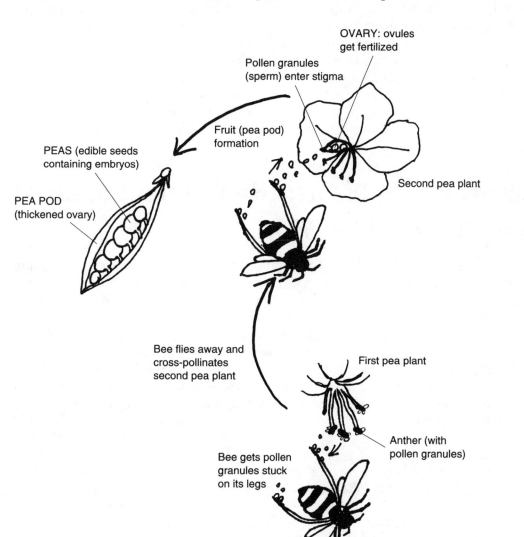

Fig. 9.6 The pea and its pod: An example of cross-pollination.

flowers. Rather, the wind blows pollen from one plant to another, or perhaps an insect, such as a bumblebee, gets a lot of pollen rubbed off onto its legs from the male anthers of one pea plant it is visiting. The bumblebee flies across to a second pea plant, which it pollinates by dropping some of the pollen granules from its legs, into the pipe-like stigma and style of a white pea flower. The pollen granules fall down and fertilize an ovule within the underlying ovary, creating a zygote, then an embryo. The embryo is contained within a seed, surrounded by protective fruit tissues. In the pea plant, for instance, the peas themselves are the mature ovules or seeds, created after cross-pollination and fertilization. And the pea pod is the casing of fruit (mature ovary with thickened walls) surrounding them. Eventually, the pea pod drops to the ground, or it is eaten by birds or animals or humans. The sweet flesh of the peas is digested, but the tough pea seeds and embryos remain unharmed. Eventually, the animal drops the seeds in its excrement, which acts as a fertilizer to help the seeds sprout or germinate into new pea plants.

Monocots versus Dicots: Two Different Kinds of Flowering Plant Embryos

We have talked about cross-pollination and fertilization of the ovules within flowering plants. These processes produce seeds containing embryos that are surrounded by fruit tissue. Such embryos first emerge from seeds as *cotyledons* (**kaht**-uh-**LEE**-duns) literally "small cups or hollows." A corn seed (kernel of corn), for example, has only one cotyledon appear as a small cup or hollow within it. A corn plant, therefore, is classified as a *monocot* (**MAHN**-uh-**kaht**), which is short for *monocotyledon* (**mahn**-uh-kaht-uh-**LEE**-dun). Corn plants, grasses, lilies, palm trees, and irises are all examples of monocots (monocotyledons), because they each have only "one or a single" (*mono-*) cotyledon (small cup or hollow) appearing as an embryo within their seeds. Such monocot embryos usually grow into mature plants having leaves with parallel-running veins and flowers with three parts.

The other major group of angiosperm embryo types are the *dicots* (**DIE**-kahts) or *dicotyledons* (die-**kaht**-uh-**LEE**-duns). This group of plants contains "two" (*di-*) small, cup-shaped embryos within its seeds. The great majority of living plant species are dicots (dicotyledons). These include peas and most other cultivated plants, as well as many types of trees. Dicots mature into plants bearing leaves with networks of veins, and flowers that split into four or five parts.

Medicinal Plants: Plants Promoting Biological Order

Throughout recorded history, human beings have exploited plants and their products not only for food but also as sources of healing for their diseases and injuries. Many of our modern drugs are potent extracts or synthetic modifications of the parts of *medicinal plants*. Consider, for example, the substance called *ephedrine* (ih-**FED**-rin). This drug was named after its original natural source – a group of gymnosperms called the *Ephedra* (ih-**FEE**-druh), literally "horsetail" plants. The Ephedra (horsetail plants) are a genus of small shrubs whose long stems are jointed and wiry, making them look like (of course) horsetails! Their leaves are flat, like scales, helping them to conserve water and adapt to life in the desert regions of Asia, Europe, and North and South America. Long ago, Native Americans ground up these shrubs and used them as a food source (flour and so-called Mormon or Mexican tea). But, even more importantly, they discovered that the Ephedra were strong medicinal plants. We now know that the chief chemical was ephedrine, which is a powerful *bronchodilator* (**brahng**-koh-**DIE**-lay-ter) – a "widener" (*dilat*) of the "bronchi" (branching lung airways). Ephedrine is thus often given to asthma patients, helping them to relax and open their airways during an attack. Further, ephedrine is a *constrictor* (kahn-**STRIK**-ter) or "narrower" of the blood vessels lining the nasal passages. Thus, it greatly reduces the swelling and inflammation of the nasal passages during hay fever attacks.

Poisonous Plants: Promoters of Disease and Disorder

Take a walk through any woods or field, and you are likely to encounter the exact opposite of the medicinal plants – the *poisonous or toxic plants*. These poisonous plants are generally classified into "Do not eat!" groups, versus "Do not touch!" groups. The "Do not eat!" category includes such species as poison hemlock, castor bean, horse chestnuts, laurel, jimsonweed, and deadly nightshade. Prominent among the "Do not touch!" group are *poison ivy*, *poison oak*, and *poison sumac* (Figure 9.7). Poison ivy and poison oak are vines that climb on trees and fence posts, whereas poison sumac is a stand-

1, B-Web

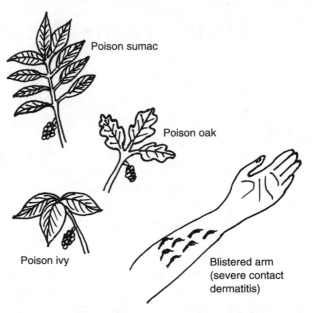

Fig. 9.7 "Do not touch!" – The poisonous trio.

1, Disorder

alone shrub-like plant. [**Study suggestion:** Look carefully at Figure 9.7. What prominent anatomic feature do all three poisonous varieties of plants have in common? This will help you recognize them, so you "Do not touch!"]

These closely related plants all contain the same toxic chemical, *urushiol* (yew-**ROO**-she-ahl). Urushiol is a toxic oil that is easily rubbed off onto the human skin when the leaves of poison ivy/oak/sumac are touched. The oil frequently creates severe *contact dermatitis* (**der**-mah-**TIE**-tis), a painful "inflammation of" (-*itis*) the "skin" (*derm*). The affected skin may become extremely red, itchy, and swollen, and drain watery fluid from blisters. Urushiol may also be a *sensitizer,* making the stricken person even more sensitive to poison ivy in the future!

Quiz

Refer to the text in this chapter if necessary. A good score is at least 8 correct answers out of these 10 questions. The answers are found in the back of this book.

1. A plant is:
   (a) A multicellular organism that develops from an embryo, contains chlorophyll, and produces energy by photosynthesis
   (b) Seldom found associated with fungi or animals
   (c) Solely engaged in only the process of photosynthesis, nothing else
   (d) A heterotrophic organism that gains most of its nutrition by feeding off the decaying bodies of others

2. The vascular plants:
   (a) Typically are without any true leaves, stems, or roots
   (b) Include the mosses and other moss-like plants
   (c) Contain no hollow ducts within their tissues
   (d) Are alternately called the tracheophytes

3. A bryophyte is technically:
   (a) A vascular plant without true leaves
   (b) A "tree moss plant"
   (c) Some kind of seedless plants without leaf veins
   (d) Related closer to the slime molds than to other plants

4. Hepatica:
   (a) Is a low-growing plant with lobed leaves resembling the liver
   (b) Can be easily compared to a muscle in the lower leg
   (c) Commonly grows to over 100 feet in height
   (d) Usually lives on dry, wind-swept mountaintops

5. The main group of seedless vascular plants:
   (a) Oak trees
   (b) Mildews and slime molds
   (c) Ferns with fronds
   (d) Gymnosperm flowers

6. The vascular plants with "naked" seeds:
   (a) Sperms
   (b) Gymnosperms
   (c) Angiosperms
   (d) Ovas

7. Pine cones produce spores by the process called:
   (a) Symbiosis
   (b) Evolutionary adaptation
   (c) Protective camouflage
   (d) Meiosis

8. A cloud of pollen is really:
    (a) Millions of microspores that have developed into male gametes
    (b) Thousands of megaspores ready to be fertilized
    (c) A pair of ovules within a female scale
    (d) A female cone that exploded accidentally

9. The male reproductive organ of a flower:
    (a) Carpel
    (b) Stamen
    (c) Stigma + style
    (d) Ovary

10. Dicots are flowering plants that:
    (a) Contain only one small cup or hollow in their seeds as a beginning embryo
    (b) Include corn plants, grasses, lilies, palm trees, and irises
    (c) Enclose a pair of hollow embryos within each of their seeds
    (d) Bloom only in the desert

The Giraffe ORDER TABLE for Chapter 9
  (Key Text Facts About Biological Order Within An Organism)

1. _____

2. _____

3. _____

4. _____

5. _____

The Dead Giraffe DISORDER TABLE for Chapter 9
(Key Text Facts About Biological Disorder Within An Organism)

1. _____

The Spider Web ORDER TABLE for Chapter 9
(Key Text Facts About Biological Order Beyond The Individual Organism)

1. _____
2. _____
3. _____
4. _____
5. _____

The Broken Spider Web DISORDER TABLE for Chapter 9
(Key Text Facts About Biological Disorder Beyond The Individual Organism)

1. _____

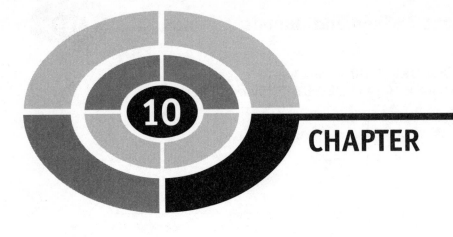

CHAPTER

Invertebrates As Special Animals: "Have You No *Spine*?"

The last few chapters have featured relatively simple organisms, such as the bacteria, slime molds, and fungi. And Chapter 9 considered plants in detail. Now, with Chapter 10, we begin to move beyond plants and simple organisms and explore the Exciting World of Animals!

You may remember (Chapter 3) that an animal is any "living, breathing" (*anima*) multicellular organism that is neither a plant nor a fungus. We further label all animals as heterotrophs, in that they must eat other organisms or organic matter to obtain energy for their survival. Finally, all animals are eukaryotes, because their cells contain nuclei.

The Invertebrates: Animals without Backbones

Chapter 3 also pointed out that there are two broad types of animals – the vertebrates (animals with backbones) and the invertebrates (animals without backbones). We are discussing the invertebrates, first, because they are the older group and the first to appear in the Fossil Record.

There are several ways in which invertebrate animals can be subdivided into different main types or groups: for example, according to whether true body tissues are present or according to the overall pattern of body form.

1, Web

PRESENCE OR LACK OF TRUE BODY TISSUES

Recall (Chapter 2) that a tissue is a collection of living cells, plus the inter-cellular material located in the spaces between them. Most invertebrates (and other animals) contain true body tissues: i.e., collections of similar cells that are specialized to perform only certain body functions. Epithelial tissues, for example, are collections of cells specialized for covering and lining body parts, while connective tissues help strap or link different body parts together.

This main group of animals are technically called the *eumetazoans* (**yew**-met-uh-**ZOH**-ahns). The eumetazoans are literally the "animals" (*-zoans*) that contain "true" (*eu-*) tissues that are formed "after" (*meta-*) repeated cell division in the embryo. Both a starfish and a lobster, for instance, contain specialized tissues playing particular roles in different parts of their bodies. Hardened epithelial tissue covers the tough, horny back of the lobster, whereas the soft muscle tissue (commonly eaten by humans) lies within its body and legs.

2, Web

The other main group of animals are the *parazoans* (**pair**-uh-**ZOH**-ahns). The parazoans are a small group of invertebrates that literally lie "beside" (*para-*) most other "animals," not quite belonging. The parazoans lack true tissues. This makes them basically different from the eumetazoan animals with tissues. The primary living examples of parazoa are the *sponges* (Figure 10.1). Sponges are members of the *Phylum Porifera* (poor-**IF**-er-ah), because their bodies are full of "pores" or holes. Sponges are usually stationary animals attached to the ocean floor. They consist of tube-like or wider vase-like colonies of cells having some specialization, but not enough to create individual tissues. Cells with flagella help sweep large quantities of seawater in through pores along the sides of the sponge. The flagellated cells are called *collar cells*. As the nutrient-rich seawater is swept in, food particles are caught within the sticky mucus in the oval collar of tiny branches located at the base of the flagellum. The particles are then ingested by phagocytosis.

3, Web

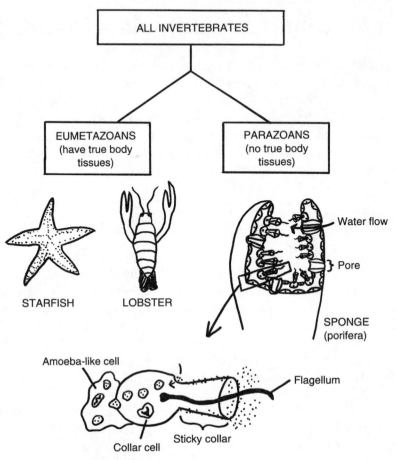

Fig. 10.1 Eumetazoans versus Parazoans: Tissues versus no tissues.

Neighboring amoeba-like cells help with the digestive process. Although the collar cells and amoeba-like cells are different, they still do not form separate types of tissues.

ORDERLY PATTERNS OF BODY FORM WITHIN INVERTEBRATES

1, Order

Besides lacking a backbone or vertebral column, the invertebrates display other orderly patterns of body form. One fundamental pattern is *symmetry* (**SIM**-et-ree). The term, symmetry, exactly translates from the Ancient Greek to mean, "the process of measuring together." In modern times, however, we are not really referring to measuring anything (as with a ruler). Rather, we are carefully looking at the shape and size of two things, compared to each other. Symmetry is said to be present, then, whenever there is a rough balance or equality of body shape and size, on either side of some dividing line. Since invertebrates lack a stiffening backbone, their bodies tend to be much more flexible than those of vertebrates. Thus, symmetry has become an important organizing influence or pattern for their survival.

There are several specific kinds of symmetry (see Figure 10.2). The most familiar to most people is *bilateral* (buy-**LAT**-er-al) or *mirror-image symmetry*. In this type of symmetry, an imaginary line subdivides the body

2, Order

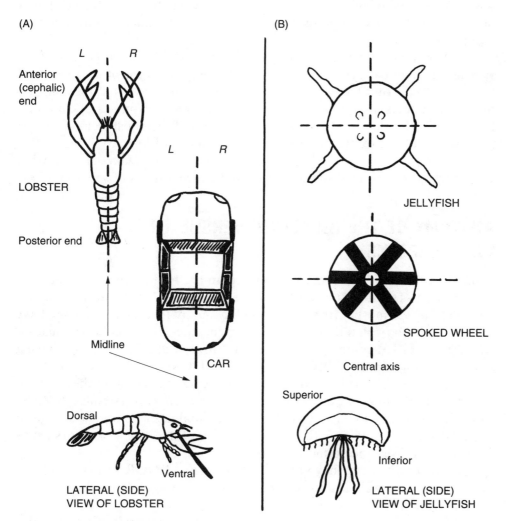

Fig. 10.2 Two kinds of symmetry in eumetazoans. (A) Bilateria (bilateral symmetry). (B) Radiolarians (radial symmetry).

into two equal halves. The right half of the body is then considered a mirror image of the left half of the body. Consider, for example, a line drawn lengthwise through the middle of a lobster (Figure 10.2, A). The lobster body often has a high degree of bilateral (mirror-image) symmetry, because the right side of the body is a mirror image of the left side, and vice versa. (By mirror image, it is meant that both sides of the body have the same shape and size, but that right and left are reversed.)

[**Study suggestion:** Get up out of your chair and go look at your reflection in a full-length mirror. To what degree does your body and its reflection show the characteristic of bilateral or mirror image symmetry?] An overhead view of an automobile also often reveals a high degree of bilateral (mirror image) symmetry.

Another important type of rough balance is called *radial* (**RAY**-dee-al) *symmetry*. In radial symmetry, there is a rough balance of various parts or "rays" (*radi*) that come out from the same center or axis. [**Study suggestion:** Picture the sun and its rays, which make a radial symmetry.] Consider, for example, the identical tendrils or arms of a jellyfish, which seem to *radiate* (**RAY**-dee-**ate**) out from a central axis in the middle of its body (Figure 10.2, B). The jellyfish, therefore, has radial symmetry. This makes it somewhat resemble a wheel with a central hub, around which a series of spokes radiate.

ANATOMY OF THE BILATERIA VERSUS THE RADIOLARIANS

Animals with a bilateral symmetry body plan are technically called the *bilateria* (**buy**-lah-**TEER**- ee-uh), or "two-sided" animals. The bilateria (bilateral animals) include most vertebrates, as well as many invertebrates. They have more to their body plan than just left and right sides. Bilateria have a head or *anterior* (an-**TEER**-ee-or) end, that lies in "front" (*anteri-*). And they have a tail or *posterior* (pahs-**TEER**-ee-or) end, that follows "behind" (*posteri-*). Since the bilateria have a "head" or *cephalic* (seh-**FAL**-ik) end to their bodies, we say that they show the characteristic called *cephalization* (**sef**-uh-luh-**ZAY**-shun). By cephalization, it is meant that an animal has a definite head end to its body, usually containing the main collection of its *sensory organs* (such as the brain and eyes and sound detectors).

3, Order

Further, the bilateria have an upper or *dorsal* (**DOOR**-sal) side in their "back" (*dors*), as well as a lower or *ventral* (**VEN**-tral) side on their "belly" (*ventr*).

Another main group of invertebrates are the *radiolarians* (**ray**-dee-oh-**LAIR**-ee-uns) – creatures with "little rays" (*radiol*) or spines projecting outward from their bodies. The jellyfish with its many radiating arms, of course, is a typical radiolarian. In these animals, there is no head or rear end, nor left or right side. They do not show the characteristic of cephalization. There is, however, both a *superior* (**soo-PEER**-e-or) portion of the animal lying "above" (*superi*) most of the body, and an *inferior* (**in-FEER**-e-or) portion lying "below" (*inferi*). (Go back and review Figure 10.2 to see these terms of relative body position.)

PRESENCE OF GERM LAYERS WITHIN THE EMBRYO

In sponges (the parazoans), there are no tissues, so the embryo does not form cell layers during its body development. In all animals except sponges, however, there are two or more *germ layers* – rings of cells within the embryo from which specialized tissues and organs are produced. The germ layers are created in several stages during the maturing of a zygote after fertilization (Figure 10.3).

4, Order

After several *cleavages* (successive cell divisions by mitosis), an eight-cell stage is followed by a *blastula* (**BLAS**-tyoo-lah), a hollow "little bladder"-like ball of cells. The central cavity within the middle of the blastula is called the *blastocele* (**BLAS**-toh-**seal**).

Following the blastula is a *gastrula* (**GAS**-true-lah). The gastrula is literally a "little stomach" (*gastrul*) or hollow ball of several layers of *germ cells*. It is these germ cells (or germ layers) that eventually give rise to the specialized tissues in the later embryo and, finally, the adult stages of life. The gastrula is created by an infolding of the layer of surface cells around the blastula. This infolding creates another cavity, called the *archenteron* (**ark-EN**-ter-ahn) – the "beginning" (*arch*) form of the "intestine" (*enteron*). The archenteron (like the mature intestine) is connected to the surface by an opening.

Around the archenteron is the *endoderm* (**EN**-doh-**derm**). The endoderm is an "inner" (*endo-*) "skin" (*derm*) of germ cells from which the lining of the intestine and interior of other major body cavities, eventually develops. An *ectoderm* (**EK**-toh-**derm**) – "outer" (*ecto-*) "skin" – covers the surface of the embryo. The ectoderm ultimately gives rise to the skin and the *central nervous systems* of many types of animals. Finally, most eumetazoans have a third germ layer, the *mesoderm* (**ME**-soh-**derm**) or "middle skin," sandwiched in between the endoderm and the ectoderm. The mesoderm forms the muscles and most other organs.

Various radiolarians besides the jellyfish – such as the *hydras* (**HIGH**-drahs), *sea anemones* (ah-**NEM**-oh-**nees**), and *coral animals* – lack a mesoderm. But they still have the other two germ layers, the endoderm and ectoderm, from which all of their adult tissues eventually develop.

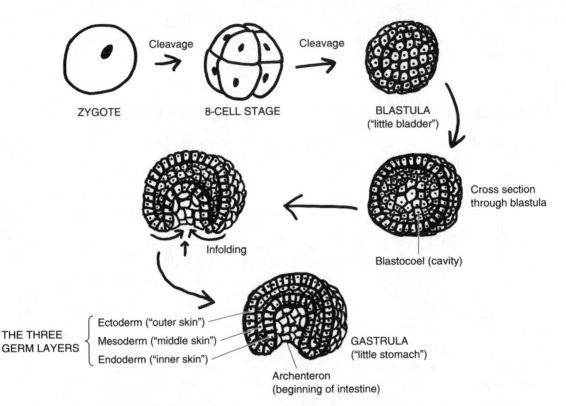

Fig. 10.3 The three germ layers in the embryo.

Coelom or No Coelom: Body Cavities in the Bilateria

Having discussed the sponges (parazoans) without tissues, as well as the radiolarians (eumetazoans with tissues and radial symmetry), we will now focus exclusively upon the bilateria. Remember that the bilateria are eumetazoans with true body tissues, as well as mature bodies that have the form of bilateral symmetry. These mature bodies grow from embryos having all three germ layers.

ACOELOMATES: NO MAIN BODY CAVITY

The previous section told us that in all animals except sponges, there is an archenteron present in the embryo. This archenteron is a hollow tube essentially representing the beginning structure of the digestive tract. All animals must have such a digestive tract, because they are heterotrophs consuming food which provides them with needed energy. Since metabolism is never 100% efficient, some of the ingested food matter is excreted through the *anus* (**AY**-nus) as feces.

A further way of classifying the bilateral invertebrates can now be considered. It is the answer to the following question: "Does the animal's body contain a *coelom* (**SEE**-loam) – central body "cavity" – around its archentereon (digestive tract), or not?"

The additional way of classifying animals thus becomes one of distinguishing the *coelomates* (**SEE**-luh-**mates**) from the *acoelomates* (**ay-SEE**-luh-mates). The coelomates, quite obviously, are those bilateral invertebrates whose bodies contain a coelom (central cavity), whereas the acoelomates have no central body cavity.

4, Web

The distinction becomes clear when one examines Figure 10.4. The acoelomates are generally considered the more primitive or most ancient organisms (according to the Fossil Record). Representative of the acoelomates are the *planaria* (plah-**NAIR**-ee-uh). The planaria are free-swimming *flatworms* that have a solid body (containing no coelom around their digestive tube). The planaria are carnivores. They catch and eat smaller animals, and feed on dead organisms in the water. The planaria have amazing powers of *regeneration*. This means that they are able to re-grow large portions of their bodies when they are cut off and removed. This trait has made them a valuable research tool for biologists seeking to learn how to promote regeneration of lost or damaged human body parts. [**Study suggestion:** Compare the name planaria with similar words like plantar and plaintain (Chapter 10). From a look at Figure 10.4 (A), after what specific characteristic is the planaria worm named?]

Midway between the acoelomates and the coelomates is a very large group of invertebrates called the *pseudocoelomates* (**SOO**-doh-**see**-luh-mates). This group is so large because it contains over 90,000 known species of *nematodes* (**NEM**-ah-toads). Nematodes are slender, "thread" (*nemat*)-"shaped" (*-ode*) worms. The nematodes (threadworms) are alternately called the *roundworms*. But their bodies tend to be narrow and cylinder-shaped, and they are often tapered at either end (Figure 10.4, B).

The nematodes (roundworms, threadworms) are classified as pseudocoelomates because their bodies contain a "false" (*pseudo-*) "cavity" (*coelom*).

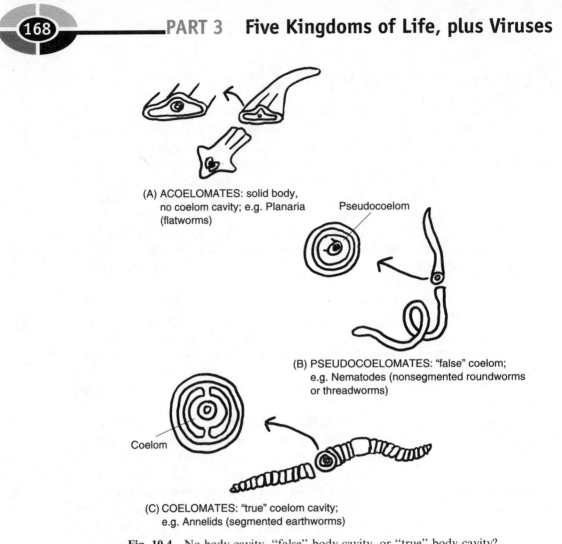

(A) ACOELOMATES: solid body, no coelom cavity; e.g. Planaria (flatworms)

Pseudocoelom

(B) PSEUDOCOELOMATES: "false" coelom; e.g. Nematodes (nonsegmented roundworms or threadworms)

Coelom

(C) COELOMATES: "true" coelom cavity; e.g. Annelids (segmented earthworms)

Fig. 10.4 No body cavity, "false" body cavity, or "true" body cavity?

These slender worms have a complete digestive tube, which extends from the mouth all the way to the anus. Their muscles run lengthwise through their entire bodies, which are *nonsegmented* (not divided into small segments). But there is no "true" coelom (cavity with an actual lining of mesoderm) between the muscles and the digestive tube. There is, instead, a pseudocoelom cavity that is filled with fluid, but is not lined with mesoderm.

Finally, we turn to the coelomates. The coelomates are commonly represented by the *annelids* (**AN**-eh-**lids**) – the worms with many "little rings" of muscles encircling their bodies. The most familiar annelid is the common earthworm (Figure 10.4, C). The earthworm is considered a *segmented worm*, due to the division of its body into numerous ring-like segments. The fluid-filled space surrounding its digestive tract is a true coelom, because it is lined by cells from the mesoderm. Earthworms are probably the most frequently studied of all coelomates in introductory biology classes. But human beings, like earthworms, are also coelomates! In both annelids (such as earthworms) and humans, the fluid within the coelom acts as a valuable shock absorber, cushioning the internal organs from blows hitting the outer body surface. It also moistens and lubricates them, reducing their friction and rubbing during body movements.

The Mollusks: "*Clam* Up, Would ya'?"

Closely related to the segmented worms is another group of coelomates, the *mollusks* (**MAHL**-usks). This phylum of mollusks consists of a huge number of more than 100,000 different species of invertebrates with "soft bodies" (*mollusc*) that are nonsegmented. Mollusks include clams, snails, oysters, squids, and octopuses. "What could such more-or-less round-shaped organisms without segments possibly have in common with the segmented annelid worms?" you might well ask. The answer is that both annelids and mollusks probably have a common ancestor in the Fossil Record, and they were the first two groups of animals to develop a true fluid-filled coelom cavity, lined with mesoderm cells.

BASIC BODY PLAN OF THE MOLLUSKS

Because they have a soft body, many kinds of mollusks are protected by a rock-hard, calcium-rich shell. Those early mollusks who developed this adaptation would obviously have enjoyed a much greater protection from hungry predators. Squids and octopuses have either a much-reduced shell, which is mostly internal, or they have lost their shells entirely.

Despite their obvious differences in degree of protection by shells, the mollusks all share a common basic body plan. The basic plan is shown in Figure 10.5. The three main parts of any mollusk are the *mantle, visceral* (**VIH**-sir-al) *mass,* and *foot.* The mantle is a "little cloak" or covering on the back of the mollusk. The covering tissue within the mantle produces the hard shell in many species. And since it is on the back or superior surface of the mollusk, the mantle functions in respiration, exchanging gases with the surrounding air or water. In hundreds of mollusk species, there is a *mantle cavity* also present below the mantle itself. The mantle cavity contains, in turn, a set of *gills*. Since the mantle is hardened by an additional shell, it has a reduced

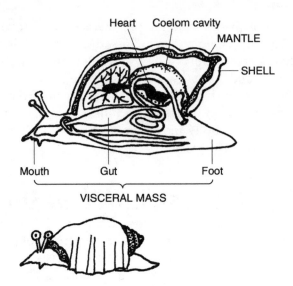

Fig. 10.5 Basic body plan of the mollusks.

surface area available for diffusion of oxygen (O_2) into the blood, and diffusion of carbon dioxide (CO_2) out. This reduced surface area available for respiration is compensated for by the gills, which assist with this critical job.

The visceral mass is the main soft body of the mollusk, which contains the major *viscera* (**VIH**-ser-**ah**) or "guts" (internal organs). The coelom cavity is also found within the visceral mass. It holds the heart of the organism. The foot of the mollusk is the inferior portion of the visceral mass, which, somewhat like a human foot, is large and fleshy and flat. The foot pushes against the ground or sea bottom, propelling the mollusk forward.

In addition to their basic body plan, many species of mollusks are classified as *bivalves* (**BUY**-valves). This is because they have "two" (*bi*) shells hinged tightly together, like a "valve." This double-valve, when opened, gives the organisms bilateral symmetry. Bivalves include such well-known invertebrates as the clams, oysters, mussels, and scallops.

Echinoderms: "Such A *Prickly Skin* May Make You A Star!"

Bivalves have bilateral (mirror-image) symmetry, while adult starfish possess radial symmetry. Various species of starfish, along with the *sea urchins* (**UR**-chins), belong to a phylum of invertebrates called the *Echinoderms* (ih-**KY**-

nuh-derms). The Echinoderms are a group of small sea animals covered by a spiny "skin" (*derm*) that forms a hard protective shell. This phylum gets its name from the numerous spines on the surface of the animals, which give them the appearance of a sea-dwelling "hedgehog" (*Echino-*). The sea urchins, in particular, look much like real "hedgehogs" (*urchins*)! The sea urchins (Figure 10.6, A) are a group of small, round Echinoderms with hard shells consisting of moveable, calcium-hardened spines. Like a hedgehog, the animal curls up and projects its sharp spines as a protective reaction when it is being attacked by a hungry predator.

5, Web

SEA URCHIN ("HEDGEHOG")   Real curled-up hedgehog

(A)

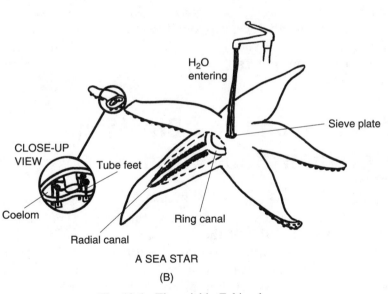

Fig. 10.6   The prickly Echinoderms.

The starfish (not being fish), are more accurately called *sea stars*. Like the sea urchins, the sea stars possess radial symmetry in their adult form. Interestingly enough, however, all the Echinoderms grow into radially symmetrical adults from *larvae* (**LAR**-vee) or immature "ghosts" (*larva*) that are bilateral and free-swimming. The conversion of immature larvae with bilateral symmetry into mature Echinoderm adults with radial symmetry apparently reflects an adaptation to a more-or-less stationary and slow-moving existence on the ocean floor.

5, Order

The sea star consists of a central body, with five arms radiating out like spokes from the hub of a wheel (Figure 10.6, B). Its inferior mouth opens into a digestive tract. Besides its prickly skin, the sea star shows another feature unique to the Echinoderms: a *water vascular* (**VAS**-kyoo-lar) *system.* This vascular system is contained within the animal's coelom (main body cavity). A small *sieve plate,* present on the superior surface of the animal, filters seawater (like a sieve with holes) just before it enters the sea star and flows down a short tube. The tube connects to a central *ring canal,* and from there into individual *radial canals,* which circulate the filtered water into each prickly arm.

Besides acting as a circulatory system for nutrients and waste products, the water vascular system serves as a *hydrostatic* (**HIGH**-druh-**stat**-ik) *skeleton.* The "water" (*hydro-*) within the vascular canals provides a "steady" (*static*) pressure. This hydrostatic pressure allows the sea star to use the suckers on its hundreds of *tube feet* to push down and slowly drag its prickly body across the ocean floor. The water-driven pressure is even great enough to allow the star to raise its arms and grasp prey, such as sponges, mollusks, and oysters.

Invertebrates Respond to "Breaking Symmetry" of Their Body Form

We have talked about the concept of Biological Order, as contrasted with Biological Disorder, as a consistent theme in this book. Symmetry, a rough balance in the shape and size of the body parts within an organism, is an important example of one kind of Biological Order. Thus, the bilateral symmetry of planaria worms, for example, as well as the radial symmetry of adult sea stars, are likewise models of biological balance and order.

1, Disorder

What happens, then, when part of a planaria flatworm is cut off? Or what happens when one of the arms of a sea star is clipped? This type of disturbance is often called *symmetry-breaking.* By breaking of symmetry, it is meant that the rough balance between the parts of an organism has been disturbed. In humans, for example, amputating, say, the left leg below the knee, creates a severe symmetry-breaking. Because the left leg is now significantly shorter than the right leg, the normal bilateral symmetry of the human body has largely been broken. Without an artificial limb to correct this imbalance, the person's ability to walk or even stand upright is severely compromised.

In sea stars and planaria flatworms, however, being relatively primitive invertebrates gives them certain advantages whenever their body symmetry is broken. Soon after a sea star loses an arm, for instance, the resulting breaking of its radial symmetry throws the animal into a state of severe imbalance as it tries to crawl over the ocean floor. This sudden introduction of great Biological Disorder strongly stimulates the cells within the stump of the missing arm to extensively grow and divide by mitosis. Due to the sea star's amazing powers of regeneration, another arm soon grows to replace the one that was torn off. Unlike humans and other vertebrates, invertebrates like the sea star have no need for crutches!

Quiz

Refer to the text in this chapter if necessary. A good score is at least 8 correct answers out of these 10 questions. The answers are found in the back of this book.

1.   The eumetazoans:
     (a) Seldom contain tissues within their bodies
     (b) Are usually incapable of mitosis
     (c) Contain true body tissues formed after cell division in the embryo
     (d) Are the main group of vertebrate organisms

2.   Symmetry:
     (a) Is a characteristic only found in the invertebrates
     (b) Provides an important example of Biological Disorder
     (c) Indicates that a rough balance of body shape and size exists on either side of some dividing line
     (d) Is frequently greatly reduced or absent in the larvae of many organisms

3.   Mirror-image symmetry:
     (a) Suggests bilateral symmetry between the right and left sides of a particular organism
     (b) Essentially explains the body form of most adult jellyfish
     (c) Is usually missing in the bodies of adult humans

(d) Involves the tricky business of organisms trying to deceive predators, by casting their reflections as "mirrors"

4. By "cephalization" in an animal, it is meant that:
   (a) The creature has only superior and inferior surfaces
   (b) The features of a radiolarian are being expressed
   (c) The animal cannot tell its right from its left
   (d) There is a definite head end to its body, where the main collection of sensory organs are located

5. Germ layers:
   (a) Represent "germs" (bacteria) that contaminate otherwise healthy cell "layers"
   (b) Are rarely observed in either vertebrate or invertebrate embryos
   (c) Arise as rings of cells within the embryo, from which specialized tissues and organs eventually develop
   (d) Occur as flat sheets of damaged cells in invertebrate adults

6. The inner skin of the embryo, from which the lining of the intestine and other major cavities develops:
   (a) Gastrula
   (b) Blastocele
   (c) Ectoderm
   (d) Endoderm

7. Coelomates:
   (a) Are the only type of organisms containing an archenteron
   (b) Are the only types of organisms having a central cavity surrounding their archenteron
   (c) Never contain an archenteron
   (d) Have solid bodies without internal cavities

8. The nematodes:
   (a) Have nonsegmented bodies, complete digestive tracts, but a pseudocoelom rather than a true coelom
   (b) Include the tapeworms, which have both coeloms and true body tissues
   (c) Are a group of solid roundworms with no internal body cavities, whatsoever
   (d) Is a small group of only a few dozen known species

9. The Echinoderms:
   (a) Seldom occur outside of dense forested environments

(b) Being so soft-bodied, have left behind virtually no traces within the ancient Fossil Record

(c) May be alternately classified as bivalves

(d) Can be nicknamed as the "hedgehogs" of the Sea!

10. Which of the following would be a good example of "breaking symmetry" of a sea urchin?

(a) Slicing the animal exactly in half along its body midline

(b) Pulling out every single one of the animal's spines

(c) Extracting all of the spines on only the left side of the urchin's body

(d) Pulling out every other spine, all over the surface of the urchin's body

The Giraffe ORDER TABLE for Chapter 10
(Key Text Facts About Biological Order Within An Organism)

1. _____

2. _____

3. _____

4. _____

5. _____

The Dead Giraffe DISORDER TABLE for Chapter 10
(Key Text Facts About Biological Disorder Within An Organism)

1. _____

The Spider Web ORDER TABLE for Chapter 10
(Key Text Facts About Biological Order Beyond The Individual Organism)

1. _____

2. _____

3. _____

4. _____

5. _____

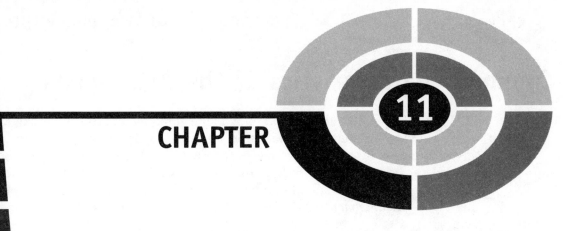

CHAPTER 11

The Arthropods: No Jointed Backbone, but "Jointed Feet"

Chapter 10 introduced us to the invertebrates – the animals "without" (in-) "jointed backbones" (vertebr). Technically, the backbones are called the *vertebrae* (**VER**-tuh-**bray**). Humans and other vertebrates contain such jointed backbones that support their body weight.

This chapter now introduces another group of invertebrates – the *arthropods* (**AR**-thruh-**pahds**). Arthropods, like other invertebrates, contain no jointed backbones. But they are set distinctly apart, however, by their "jointed" (*arthro*-) "feet" (*pod*)!

1, Web

Common Characteristics of the Arthropods

As Figure 11.1 shows, the general body plan shared by all arthropods can be represented by the anatomy of a lobster. The arthropods are a huge phylum of invertebrates with bilateral symmetry whose bodies are clearly divided into jointed segments, and whose legs are hollow and also divided into jointed segments. Possessing bilateral symmetry, the arthropods have pairs of matching right–left *antennae* (an-**TEN**-ee) or feelers, legs, and in many insects, wings.

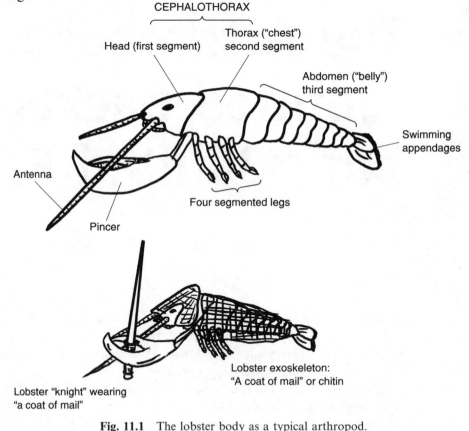

CEPHALOTHORAX

Head (first segment)

Thorax ("chest") second segment

Abdomen ("belly") third segment

Swimming appendages

Antenna

Four segmented legs

Pincer

Lobster exoskeleton: "A coat of mail" or chitin

Lobster "knight" wearing "a coat of mail"

Fig. 11.1 The lobster body as a typical arthropod.

1, Order

THE SEGMENTED BODY

The lobster is covered by a hard *exoskeleton* (**eks**-oh-**SKEL**-eh-ten) – an "outer" (*exo-*) "hard dried body" (*skeleton*). Since the exoskeleton is on the surface of the animal, it is also called the *cuticle* (**KYOO**-tih-kl) or "little skin." The exoskeleton (cuticle) can even be thought of as a "coat of mail," a type of armor! Why is this? The reason is that the lobster exoskeleton contains large amounts of *chitin* (**KY**-tin). Chitin also covers the surfaces of crabs, beetles, and crickets. Chitin is a tough, horny substance providing protection to the invertebrate body, much as if the lobster were an ocean-dwelling knight wearing a "coat of mail" (*chitin*)!

The chitin-armored lobster body has two pairs of antennae (short and long), two feeding mouthparts, as well as two large pincers, attached to either side of the head. A *thorax* (**THOH**-racks) or "chest" piece forms a second segment behind the head. The head and thorax are often lumped together by biologists and named as a single large segment called the *cephalothorax* (**sef**-uh-low-**THOR**-aks) or "head" (*cephal*)-and-"chest" piece. Four hollow, segmented legs are attached to either side of the cephalothorax and allow the lobster to easily walk over the ocean floor.

Behind the thorax is the third body segment – a much-longer *abdomen* (**AB**-duh-mun). The abdomen is literally the long "belly" of the beast. But for we lobster-eating humans, it is usually just considered the tail! The exoskeleton protecting the abdomen is subdivided into a number of hinged plates, allowing the lobster to use its *swimming appendages* (uh-**PEN**-dih-jes) or "hangers-on" to scoot its body rapidly along.

AN OPEN-ENDED CIRCULATORY SYSTEM

Arthropods do have hearts, but these hearts pump a fluid called *hemolymph* (**HE**-moh-**limpf**), rather than blood. The circulatory system is an open one. This means that the heart pumps the hemolymph into vessels and sinuses which are not closed off. Nevertheless, the hemolymph still provides the body tissues with an adequate supply of oxygen and other nutrients.

A DEPENDENCE UPON MOLTING

Because many arthropods have a rigid exoskeleton, yet they continue to grow in size throughout most of adulthood, they are forced to periodically *molt* (**MOHLT**) or shed. After the exoskeleton is molted (shed), the lobster or other arthropod is temporarily soft-bodied and vulnerable to hungry predators. But the head of the lobster bears two eyes, each perched upon a movable stalk, that lets it readily see approaching predators (and hopefully escape being eaten!).

Major Types of Arthropods

There are several major types or categories of arthropods. The first question a person might ask, however, is: "The arthropods are all invertebrates. So, why weren't they included within Chapter 10, which described the invertebrates?" [**Study suggestion:** Take a careful look at Figure 11.2. What particular feature really sticks out at you? How do you think this feature explains why arthropods are placed in a separate chapter of their own?]

2, Web

Using a traditional approach to taxonomy, there is a single phylum *Arthropoda* (ar-**THRAH**-pah-da), subdivided into five smaller classes. These are the *Crustaceans* (krus-**TAY**-shuns), *Arachnids* (ah-**RACK**-nids), *Chilopods* (**KY**-luh-pods), *Diplopods* (**DIP**-luh-pods), and *Insects*. In Figure 11.2, the relative size of each oval shape representing these arthropod classes visually pictures the number of known species belonging to each one.

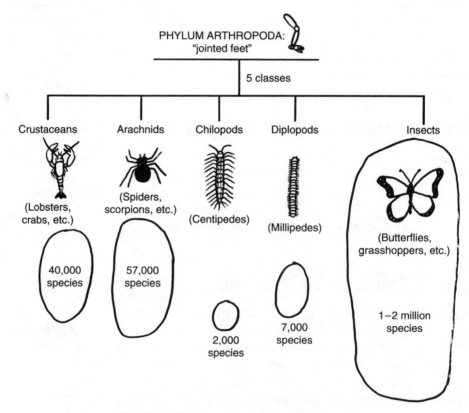

Fig. 11.2   The five major classes of arthropods.

ANCIENT TRILOBITES AND THE MODERN CRUSTACEANS

Let us begin our examination of the major arthropod classes with the *crustaceans* – invertebrates whose bodies are surrounded by hard "shells" (*crustace*). In the modern world, the crustaceans include crabs, lobsters, crayfish, and shrimp. Besides their hard shells (rigid exoskeletons), these crustaceans have jointed bodies and legs, as well as *gills* that allow them to exchange their hemolymph gases with the external gases in the surrounding water.

There are now about 40,000 known species of crustaceans, most of them living within either freshwater or saltwater environments. Yet, biologists have repeatedly observed an early ancestor (long extinct) of the modern crustaceans in the Fossil Record. This extinct group of arthropods is the *trilobites* (**TRY**-loh-bites). The trilobites consisted of about 4,000 species, and they were very abundant in the sea during the Paleozoic Era (Chapter 3). The trilobites had bodies comprised of "three" (*tri-*) vertical "lobes" (*lob*) subdivided into many horizontal segments (Figure 11.3). Sprouting from either side of the body segments were numerous jointed legs. But unlike the modern lobster and other crustacea, these ancient marine invertebrates had appendages running along either side of their body that showed very little difference or specialization. For instance, they lacked the protective pincers of present-day crabs and lobsters.

The trilobites rather suddenly became extinct about 250 million years ago, near the end of the Paleozoic Era (shortly before the dinosaurs appeared

Trilobite ("three-lobes" with "segments") fossil

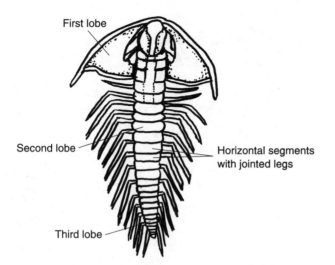

First lobe

Second lobe

Horizontal segments
with jointed legs

Third lobe

Fig. 11.3   The three-lobed body of the ancient trilobite.

3, Web

during the Age of Reptiles). Because they never reappear at later stages in the Fossil Record, the limited time period of trilobite existence strongly supports the concept that once a group of animals becomes extinct, they generally stay extinct! There is a definite Biological Order or recognizable sequence of body forms and patterns that seem to precede or follow one another during the long time-scale of evolution.

THE ARACHNIDS: SPIDERS AND THEIR RELATIVES

The *arachnids* are literally "spiders or webs" (*arachn*). Besides the spiders, however, the arachnids include scorpions, ticks, and mites as well. The arachnid group includes about 57,000 species. The basic anatomy of the arachnids is displayed within Figure 11.4. The spider body has two segments – head and thorax (cephalothorax). Situated farthest in the front of the head are the *pedipalps* (**PED**-uh-palps) and *chelicera* (kuh-**LIS**-er-ah). The pedipalps are a pair of "feeling" (*palp*) "feet" (*pedi*) – small foot-like appendages that help sense when prey is present. The chelicera or "claws" (*cheli*) then come into play. The two chelicera serve much like fangs, injecting poison from a gland just below the four eyes. The chelicera inject digestive juices, as well as poison, into captured prey, which softens the attacked body

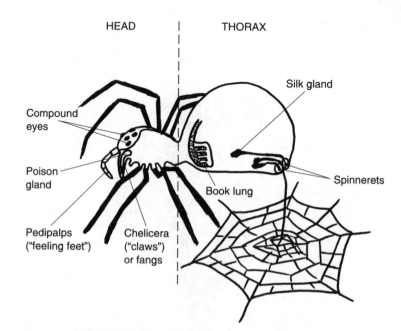

HEAD | THORAX

Compound eyes

Poison gland

Pedipalps ("feeling feet")

Chelicera ("claws") or fangs

Silk gland

Spinnerets

Book lung

Fig. 11.4 The amazing body of a spider.

as well as killing it. The spider finally sucks the liquefied food into its digestive system.

Along each side of the cephalothorax, spiders have four pairs of walking legs. Other special characteristics include a stack of *book lungs*, as well as a *silk gland* with *spinnerets* (**SPIN**-uh-rets) – small organs used for spinning webs.

CENTIPEDES AND MILLIPEDES: "SO MANY LEGS!"

The Chilopod class literally consists of arthropods with "lip" (*chil*) "feet" (*pod*), obviously named for their appearance. There are approximately 2,000 species in this group. The main Chilopods are the *centipedes* (**SEN**-tuh-peeds) or organisms with a "hundred" (*centi-*) "feet" (*ped*). Centipedes are thin, worm-like arthropods with segmented bodies having many pairs of legs (but certainly not 100!). The centipede has very long antennae and poison claws that catch smaller insects to eat (Figure 11.5, A).

The Diplopod class includes those creatures with "double" (*dipl*) "feet" (*pod*). About 7,000 species of Diplopods exist. True to their Class name, each body segment bears two legs with feet. The primary Diplopods are the *millipedes* (**MIH**-luh-**peeds**) – translated to mean arthropods with a "thousand" (*milli-*) "feet." The millipede body, of course, has far fewer than 1,000 feet (Figure 11.5, B)! Millipedes feed on moss, decaying leaves, and a variety of other types of plant matter.

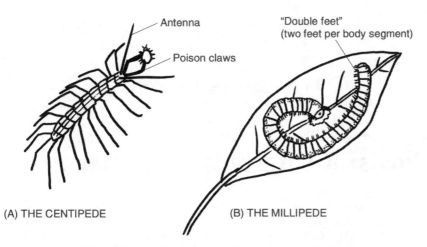

(A) THE CENTIPEDE   (B) THE MILLIPEDE

Fig. 11.5   Centipedes and millipedes: So many legs!

INSECTS: THE SWARMING HERD

2, Order

Of all the arthropods, the insects represent by far the hugest number of species – at least one million known (and maybe another million or so insect species still undiscovered, mainly within tropical forests)! This means that there are more species of insects than species of all other types of animals, combined!

Yet, despite their overwhelming numbers and diversity, insects share a common body plan (Figure 11.6). The word *insect* actually comes from the Latin for "cut" (*sect*) "into" (*in-*). The reason for this name is that the body of most insects is "cut" (subdivided) into three major segments. Using the example of a grasshopper, there is a head, a thorax, and an abdomen. The insect head bears two antennae and a pair of compound eyes. In their adult stages, many insects have three pairs of jointed legs and either one or two pairs of wings, all attached to the thorax.

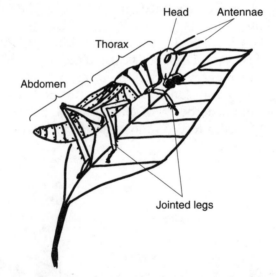

Fig. 11.6   The grasshopper: An insect body plan.

Metamorphosis: A Caterpillar Becomes A Butterfly

No doubt you have seen a humble caterpillar crawling slowly across a leaf. But hidden from your gaze, at the end of summer the caterpillar begins to undergo a dramatic *metamorphosis* (**met**-uh-**MOR**-fuh-sis) – a "changing over" (*meta-*) of its body shape or "form" (*morph*).

In most cases, metamorphosis is a mechanism whereby an embryonic stage of an animal (such as an insect) changes into a strikingly different adult body form. One of the most beautiful examples of metamorphosis is the transformation of a caterpillar into a gracious, delicate-winged butterfly. Some of the major steps in this process are shown in Figure 11.7 for the well-known *monarch* (**MAHN**-ark) butterfly. Like a king or queen or some other type of "ruler-alone" (*monarch*) in the world, the adult monarch butterfly stands alone in its striking orange-and-black body pattern.

3, Order

Larva (caterpillar on milkweed)

End of growing stage

Larva attaches

Metamorphosis begins

Differentiation

Pupa (chrysalis) inside cocoon

Adult emerges

Fig. 11.7   Caterpillar into monarch: a striking metamorphosis.

The process begins with a larva (caterpillar) stage. The black-striped monarch caterpillar (larva) likes to feed on the *milkweed* plant, named for its white, milky juice. The caterpillar just keeps gorging itself on milkweed leaves and growing throughout the summer, molting its skin several times. Finally, at the end of the growing stage, the caterpillar larva firmly attaches itself to a branch. Here it molts several more times, then encases itself within a *cocoon* (kuh-**KOON**). A cocoon is an external "shell or husk" of silky material that caterpillars spin around themselves during their preparation for metamorphosis.

When the larva is encased within a tough, protective cocoon, its new stage of development is called the *pupa* (**PYOO**-puh) or *chrysalis* (**KRIS**-uh-lis). In Latin, chrysalis literally means "golden pupa of a butterfly." Likewise, pupa means "girl or doll." So, the pupa (chrysalis) is poetically described as a "golden girl or doll" that will eventually mature into a queen-like, elegant butterfly! Perhaps the reference to "golden" is due to the rather shiny appearance of the outer wall of the cocoon enveloping the chrysalis (pupa).

4, Order

Within the pupa, a precisely timed genetic program is now turned on. The tissues of the larva are broken down, then replaced by other cells that undergo mitosis and *differentiation* (**dif**-uh-**ren**-she-**AY**-shun) – "the process of becoming different" or specialized. Certain body cells, for example, migrate to the sides of the pupa in a bilaterally symmetrical manner. The cells in these locations then *differentiate* (**dif**-uh-**REN**-she-ate) into the highly specialized structures of the developing butterfly wings.

Soon after the arrival of spring, the adult begins to emerge from the cocoon. At first, the wings are flat and wet and pressed against the sides of the butterfly body. But a pumping process pushes fluid out into the veins of the wings, stiffening and opening them. The result is a beautiful, orange, black-striped, bilaterally symmetrical monarch butterfly – one of the most impressive displays of Biological Order known to humankind!

Insects: Inducers of Biological Order and Disorder In the Environment

4, Web

Since there are more species of insects than all other types of animals combined, a convincing argument can be made that insects (not human beings) are the real "monarchs" (rulers) of this planet! Because of their sheer numbers, then, the World of Insects exerts a critical influence upon both Biological Order, and Biological Disorder, within the ecosystem of the external environment. Insects are important biological *inducers* (in-**DEW**-surs) – "leaders into" various states of environmental change.

BUSY POLLINATORS OF FLOWERS

Various insects visit flowers to feed upon their *nectar* (**NEK**-tar). Nectar is named for the "drink of the gods" in Greek and Roman mythology. The connection is probably the fact that nectar is sweet and filled with highly nutritious plant sugars. And busy bees gather this sweet nectar and transform it into an even-sweeter honey. Bees and butterflies visit flowers and feed on their nectar during the day, while moths show up at night.

During day-or-night visitations, these insects help induce pollination. You may recall (Chapter 9) that pollination is the process of spreading pollen grains from male plants to female plants, so the females can be fertilized. Without insect-induced pollination, many green plants would eventually die, and with it, most photosynthesis occurring on the Earth. Therefore, insect-induced pollination of flowering plants is a critical factor in the Web of Life.

5, Web

THE MENACING SWARMS

Just as some insects are busy inducers of pollination and promoters of Biological Order, others attack human crops in menacing *swarms*. A swarm is a large group of related insects that fly or move together. Communication among insects in the swarm may be highly orderly and efficient, allowing the huge group to carry out important tasks. A swarm of bees, for instance, may move as a group of thousands from a damaged hive to a safer area, and build another hive. The new geographic area with the hive will benefit from an increased pollination of its flowers.

But a swarm of *locusts* (**LOW**-kusts) is quite another matter! A locust is a type of grasshopper with short antennae. Due to their amazing powers of rapid reproduction, swarms of locusts can darken the sky as they fly. Wherever they land as a group of billions, whole tracts of fields and forests are quickly stripped of their leaves. The entire ecosystem in a particular area can be effectively destroyed, depriving the region of the benefits of life-giving photosynthesis. Since the days of Ancient Egypt, farmers have been faced with the difficult task of repairing the devastation of Biological Disorder wreaked by insect swarms.

1, B-Web

Quiz

Refer to the text in this chapter if necessary. A good score is at least 8 correct answers out of these 10 questions. The answers are found in the back of this book.

1. The arthropods are distinguished from other invertebrates by their:
   (a) Lack of jointed backbones
   (b) Presence of distinctly jointed legs and feet
   (c) Ability to reproduce asexually
   (d) Tendency to crawl slowly over the ground

2. By referring to the cephalothorax, it is meant that:
   (a) The arthropod body is thoroughly segmented
   (b) Invertebrates lack brain tissue
   (c) All external attachments or appendages are absent
   (d) The head and chest segments have been lumped together

3. The lobster can just keep growing bigger and bigger, for as long as it lives! "So," you may well ask, "why doesn't the lobster rip its own skin?"
   (a) The lobster doesn't have any covering or skin!
   (b) There is an intermittent period of molting
   (c) The lobster body does rip out of its skin, and then it quickly dies
   (d) A rapid healing process takes place

4. The trilobites are most accurately described as:
   (a) Primitive, insect-like creatures having two body segments
   (b) Arthropods without jointed legs
   (c) A long-extinct ancestor of the modern crustaceans
   (d) A vanished group of extremely overdeveloped mollusks

5. As a group, the arachnids:
   (a) Consist solely of spiders and their webs
   (b) Include practically all types of insects
   (c) Generally have bodies with eight or more segments
   (d) Involve scorpions, ticks, mites, and spiders

6. As a group, insects:
   (a) Have bodies literally "cut into" three major segments
   (b) Use eight hollow legs to move around
   (c) Enjoy extremely long life spans, but infrequent reproduction
   (d) Fly with at least 5 pairs of wings

7. Metamorphosis:
    (a) Generally involves the shrinkage of an adult back into a larva
    (b) Is a mechanism for fluid transport (similar to diffusion)
    (c) Always proceeds from one body form to another
    (d) Generally occurs without any need for differentiation

8. "Insects (not human beings) are the real monarchs of Planet Earth!"
    Evidence supporting this statement is the fact that:
    (a) Insects have far more "natural intelligence" than people
    (b) Insecticides often poison humans, as well as insects
    (c) Monarchs are the prettiest butterflies in the world!
    (d) There are far more species of insects than any other type of animal

9. Bees can be considered inducers of Biological Order in that:
    (a) Pollination resulting from their feeding visits for nectar allows
        complex patterns of plant life to grow and thrive
    (b) They often live in highly aggressive colonies that fiercely attack
        hive invaders
    (c) Flying only during daylight, they do not disturb dark-loving species
    (d) Their attractive yellow-and-black body pattern stimulates other
        organisms to adopt colorful forms of their own

10. Swarms of insects:
    (a) Always attack and destroy whatever lies in their path
    (b) Only occur in far northerly regions of the planet
    (c) May involve highly orderly and efficient patterns of
        communication and coordination
    (d) Show no interactions among their members, whatsoever

The Giraffe ORDER TABLE for Chapter 11
   (Key Text Facts About Biological Order Within An Organism)

1. _____

2. _____

3. _____

4. _____

The Spider Web ORDER TABLE for Chapter 11
  (Key Text Facts About Biological Order Beyond The Individual Organism)

1. _____

2. _____

3. _____

4. _____

5. _____

The Broken Spider Web DISORDER TABLE for Chapter 11
  (Key Text Facts About Biological Disorder Beyond The Individual Organism)

1. _____

CHAPTER 12

The Chordata: Animals with a "Chord" in Their "Back"

In the last two chapters (10 and 11), we were discussing the invertebrates. By definition, these are animals without any linked bones (called vertebrae) within their backs. You may remember (Chapter 3) that jellyfish, corals, sea worms, and other multicellular invertebrates were probably among the first animals to appear during evolution. Only within the last 500 million years or so have there been any *chordates* (**KOR**-dates) – animals with some kind of "chord" or "cord" present to stiffen their backs. It is this later group, called the *Phylum Chordata* (kor-**DAY**-tuh), to which we now turn with interest.

1, Web

The Notochord: Ancient Forerunner of the Vertebral Column

It is very reasonable for you, the reader, to ask, "Just what is meant by having a chord or cord in your back?" This concept is best explained by looking at Figure 12.1, which provides the general body plan of a primitive chordate. This figure shows the four general characteristics of all chordates:

1. The presence of a slender *notochord* (**NO**-tuh-kord) running down the back during at least part of the life cycle;
2. A hollow *nerve cord* that lies immediately dorsal to the notochord;
3. *Gill slits* along each side of the head; and
4. A *post-anal* (**AY**-nul) *tail* that is very muscular.

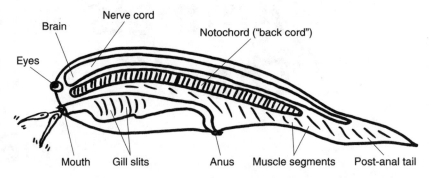

Fig. 12.1 The general body plan of a primitive chordate.

1, Order

The notochord is the key anatomic feature. It is a long, narrow structure in the "back" (*noto-*) that looks like a thin "cord" (*chord*) or rod. The notochord's chief function is supporting and stiffening the body. All chordates have a notochord in their bodies during at least one stage of their development. And it is this notochord that can serve as the basis for development of a full *vertebral column* or "jointed backbone."

The Phylum Chordata and Its Groups

The Phylum Chordata can be subdivided into a number of different groups. The major classifying factor is the answer to the following question: "Does the particular chordate species being considered keep a notochord in its back throughout its lifetime, or does it replace the notochord with a full vertebral column after it has passed through the embryo stage?''

THE SEA SQUIRTS AND LANCELETS: NO ''JOINTED BACKBONE''

Figure 12.2 reveals that two of the major groups of chordates keep a notochord in their back throughout their lives. These two groups are called the *Urochordates* (**YOUR**-oh-**kor**-dates) and the *Cephalochordates* (**SEF**-uh-low-**kor**-dates).

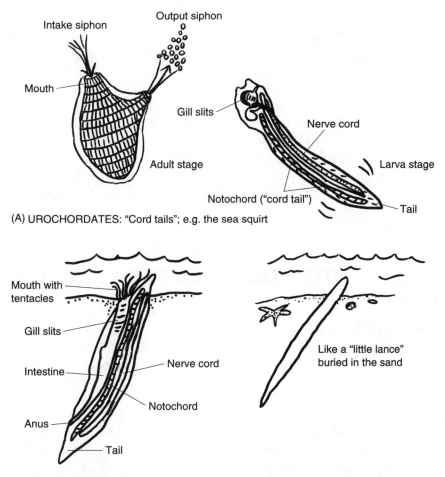

(A) UROCHORDATES: "Cord tails"; e.g. the sea squirt

(B) CEPHALOCHORDATES: "Cord-heads"; e.g. the lancelet

Fig. 12.2 Sea squirts and lancelets: Chordates without spines.

The Urochordates are animals that literally have a "cord" (*chord*) within their "tail" (*uro-*). By this it is meant that these creatures have a swimming larval stage with a notochord in its tail section. The main examples of Urochordates are the *sea squirts*. Interestingly enough, the adult stage of the sea squirt has a body shaped like a "U," and it remains anchored to a rock at the bottom of the sea. At one end lies an *intake siphon*, which sucks in seawater and filters out plankton and other tiny creatures during feeding. And at the other end lies an *output siphon*, which squirts or shoots out a jet of water whenever the animal is bothered by a predator.

The Cephalochordates are literally a group of "cord-heads" (*cephalo-*)! The implication of this odd name is displayed in Figure 12.2 (B) for the *lancelets* (**LANS**-lits). The lancelets are slender, fish-like marine animals that lie partially buried in the sand under shallow water. Their thin, tapered body is pointed at both ends, making it look much like a small spear or "little lance." Because it has no skull, the lancelet doesn't have a brain, either! Thus, the notochord extends all the way up into its head, officially making it a "cord-head" (Cephalochordate). The lancelet is a type of *suspension feeder*, meaning that it feeds on small particles suspended in seawater, which it draws into its mouth with the help of waving, hair-like *tentacles* (**TEN**-tuh-kuls).

2, Web

Both sea squirts and lancelets have notochords in their backs during at least part of their life cycles, but they are still officially invertebrates. Many biologists think that they are important evolutionary "bridging species" between the other invertebrates (which don't even have a notochord) and the true vertebrates (whose primitive notochord has been replaced by a bony vertebral column).

THE VERTEBRATES – CHORDATES WITH A "JOINTED BACKBONE"

3, Web

In contrast to the Urochordates and Cephalochordates, the *Vertebrata* (**ver-teh-BRA**-tuh) or vertebrates are chordates with both a jointed (segmented) backbone and a *brain case* or *cranium* (**KRAY**-nee-um) (see Figure 12.3). You may recall (Chapter 11) that the "jointed backbones" are technically called the vertebrae. The cranium is just a formal name for the "skull" (*crani*) "present" (-um) at the top of the vertebral column. Owing to their high degree of cephalization (concentration of sensory, motor, and other nerve functions within the head region), vertebrates have a brain in their cranium. The vertebrates generally have a notochord only as part of their embryonic stage of development. As their bodies mature, the notochord is replaced by a vertebral column (linked series of jointed vertebrae).

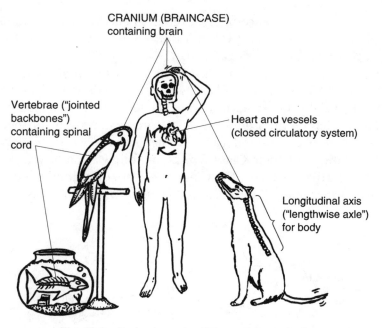

CRANIUM (BRAINCASE)
containing brain

Vertebrae ("jointed
backbones")
containing spinal
cord

Heart and vessels
(closed circulatory system)

Longitudinal axis
("lengthwise axle")
for body

Fig. 12.3 Basic elements of the vertebrate body.

The dorsal nerve cord, now protected by a vertebral column, is called the *spinal cord*. The skull (brain case or cranium), vertebral column, and ribs make up what is called the *axial* (**AX**-ee-ul) *skeleton*. The reason for this name is the pattern set by the bones or *cartilage* (**KAR**-tih-**lj**), commonly known as "gristle," which house the brain and spinal cord. This ordered bone/cartilage pattern forms a *longitudinal* (**long**-jih-**TWO**-duh-nal) *axis*, or "lengthwise axle," around which the vertebrate body can turn or pivot.

2, Order

Finally, vertebrates have a *closed circulatory system*. The heart serves as a pump, which sends the blood coursing out through the entire body. After the body tissues are supplied with oxygen, glucose, and other nutrients, the blood (now filled with tissue waste products) returns back to the heart through a closed loop of vessels.

Wide Diversity in Backboned Creatures: Eight Different Classes of Vertebrates

Although all vertebrates share a common core of characteristics, they still show a very wide degree of diversity or differences among them. This diversity is reflected in the existence of eight different classes of vertebrates. These are named the *Class Agnatha* (**AG**-nuh-thuh), *Class Placodermi* (**PLAK**-uh-**derm**-ee), *Class Chondrichthyes* (kahn-**DRIK**-thees), *Class Osteichthyes* (**ahs**-tee-**IK**-thee-eez), *Class Amphibia* (am-**FIB**-ee-ah), *Class Reptilia* (rep-**TIL**-ee-ah), *Class Aves* (**AY**-veez), and *Class Mammalia* (mah-**MAY**-lee-uh). [**Study suggestion:** The eight different vertebrate classes are each labeled with brief representative pictures in Figure 12.4. Before reading any farther in this chapter, see if you can match each of the pictures with the appropriate English translations of their Class names, as follows: the Class of "birds;" the Class of "cartilage-fishes;" the Class of "livers of a double life {on both the land and in the water};" the Class of "jawless" fishes; the Class of animals

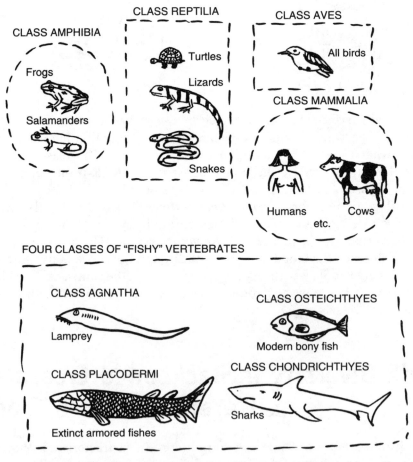

Fig. 12.4 The eight major classes of vertebrates.

with "breasts;" the Class of "bony fishes;" the Class of low-"crawling" animals; and the Class of fishes with "flat-surfaced skin."]

CLASS AGNATHA: FISH-LIKE VERTEBRATES "WITHOUT JAWS"

The members of Class Agnatha, fish-like creatures "without" (*a-*) "jaws" (*gnath*), were probably among the very first vertebrates to appear on the Earth. Chapter 3 told us that the vertebrates first appeared about 500 million years ago, during the Paleozoic or "Ancient Life" era. It was also during this era that the first amphibians and reptiles evolved.

Class Agnatha, now including about 60 species, mainly consists of the *lampreys* (**LAM**-prees) or "rock-lickers"! Lampreys, of course, do not actually lick rocks! Rather, they are slender, eel-shaped vertebrates with a large, round mouth that attaches to fish and sucks out their body fluids. Lampreys have gill slits, like a fish, but they are not generally considered a type of true fish, since they lack both a hinged jaw, as well as paired fins. (A fish in general is a vertebrate that lives in the water, has a body covered with scales, has a hinged jaw, breathes with use of gills, and has paired fins for swimming.) Due to their primitive characteristics, the modern lampreys probably resemble the first jawless vertebrates in many ways.

CLASS PLACODERMI: EXTINCT ARMORED FISHES

The Class Placodermi consisted of a group of jawed fishes called *placoderms* (**PLAK**-uh-derms) that are now all extinct! The placoderms were armored fishes with tough, "flat-surfaced" (*placo-*) "skin" (*derm*).

1, B-Web

CLASS CHONDRICHTHYES: FISH WITH SKELETONS OF CARTILAGE

The Class Chondrichthyes is technically the group of "fishes" (*ichthy*) with skeletons made of "cartilage" (*chondr*). This group chiefly includes the sharks and sea rays. Even though their skeleton is made of relatively soft cartilage, the sharks have very hard teeth that are bony. These *cartilaginous* (**kar**-tih-**LAJ**-ih-nus) fishes all have hinged jaws and paired fins.

CLASS OSTEICHTHYES: FISH WITH SKELETONS OF BONE

The Class Osteichthyes involves "fishes" (*ichthy*) whose skeletons are composed of "bone" (*oste*). This familiar Class of bony fishes (approximately 30,000 known species) is the largest group of existing vertebrates. Most of those we frequently encounter (such as perch, bass, trout, and tuna) are classified as the *ray-finned fishes*, due to the flexible ribbing of *rays* or ridges visible within their fins (see Figure 12.5).

There are several other unique features occurring in bony fishes:

1. They have a slimy skin. Coated with a secretion of *mucus* (**MEW**-kus) or "slime," the bony fishes can easily glide through the water, suffering very little friction as they move.

2. Their gills have an *operculum* (oh-**PER**-kyuh-lum) – an external "covering or lid" (*opercul*). Bony fish are able to breathe while they remain still in the water, because they take oxygen-containing water in through their mouth, and then push it out between their gills, due to the flapping movements of the operculum.

3. Their bodies contain a *swim bladder*. Unlike sharks, bony fishes have a swim bladder that they can inflate with air, allowing them to remain nearly motionless in the water without sinking. (Sharks, in contrast, have to keep constantly swimming in order to keep themselves from sinking!)

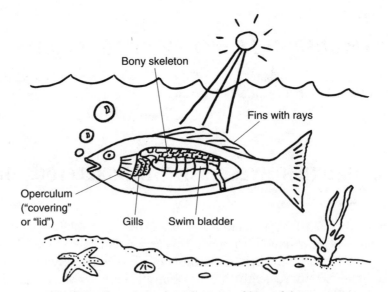

Fig. 12.5 Some important features of bony fish anatomy.

CLASS AMPHIBIA: LIVERS OF A "DOUBLE LIFE"

After the first fish appeared in the waters of the Paleozoic ("Ancient Life") Era, it was not very long before they were joined by the first amphibians. Unlike fish, which only have fins, amphibians are classified as *tetrapods* (**TET**-ruh-pods) – animals with "four" (*tetra-*) "feet" (*pods*).

Between about 350 and 400 million years ago, the Fossil Record provides evidence that there were tetrapod fishes – fish that had four foot-like attachments to their skeletons. It is from these tetrapod fishes that the first amphibians likely evolved, crawling out as pioneers from one life in the water to another on the land. Hence, the word, amphibian, literally means "liver of a double life."

The main members of the Class Amphibia are the salamanders, frogs, toads, and newts. Perhaps most reflective of the "double life" of the amphibians is the life cycle of a frog. Being hatched from eggs in the water, the frog starts out life as a legless tadpole, propelled by a muscular tail. And, like a fish, it has internal gills. As time passes, however, a dramatic metamorphosis occurs. The tail and gills are progressively *resorbed* (ree-**SORBD**) – literally "sucked in" (*resorb*) to the body, then disappear. Four legs finally develop, and the tail-less frog form at last hops out onto the land.

CLASS REPTILIA: "CRAWLERS" WITH BACKBONES

To be sure, the amphibians do more than their fair share of crawling (as well as hopping around)! But the group as a whole is named for the "double lives" its members lead – on both the land and in the water.

Amphibians have moist skins, not covered by scales. Therefore, they must frequently return to a watery environment, or else suffer the deadly consequences of tissue *dehydration* (**dee**-high-**DRAY**-shun). This term literally means "the process of" (*-tion*) losing "water" (*hydr*) "from" (*de-*) the body.

Reptiles, however, are much more adapted to life on dry land. Their skin is covered by scales containing *keratin* (**CARE**-uh-tin) or "horn" (*kerat*) "substance" (*-in*). Keratin is a family of tough, waterproof proteins found in the skin, hair, claws, and horns of various animals. Hence, one might well nickname the reptiles as the "*keratinized* (car-**AT**-uh-nized) crawlers"! This group includes the turtles, lizards, alligators, and crocodiles, which all move about by crawling. But it also involves the snakes, which have lost their limbs during the course of evolution, and so are now forced to slither!

Reptiles breathe through lungs rather than gills, and their eggs are encased within rubbery, waterproof shells. And within each of these waterproofed

3, Order

4, Order

eggs, the reptile embryo undergoes its development inside an *amnion* (**AM**-nee-un), much like a "little lamb." An amnion is a fluid-filled sac that encloses the embryos of the so-called higher vertebrates – the reptiles, birds, and mammals (Figure 12.6).

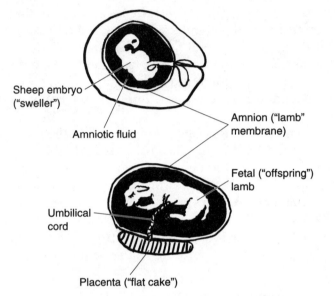

Sheep embryo ("sweller")

Amnion ("lamb" membrane)

Amniotic fluid

Fetal ("offspring") lamb

Umbilical cord

Placenta ("flat cake")

Fig. 12.6   The amnion: A "little lamb" around the embryo.

Because their embryos are encased within the amnions of waterproof eggs, they breathe through lungs, and their bodies are covered by keratin-stuffed scales, it is not so surprising that the reptiles followed the amphibians out of the water, and onto the land. They are much more protected against dehydration! Hence, they can stay on the land much longer than amphibians. (Alligators and crocodiles, of course, love the water!)

Recall (Chapter 3) that the Mesozoic Era (200–65 million years ago) was called the Age of Reptiles. The dinosaurs and many other long-extinct species of reptiles flourished and dominated the surface of the planet during this time. In other words, the entire Earth was crowded and swarming with "creepy-crawlers"!

CLASS AVES: THEY'RE JUST "FOR THE BIRDS!"

The Fossil Record strongly suggests that birds evolved from reptiles. A bird is a *homeothermic* (**HOH**-mee-uh-**ther**-mik) vertebrate that maintains its body "heat" (*therm*) about the "same" (*homeo*), has wings, feathers, two legs, a bill or a beak, and lays eggs. Becoming homeothermic, and thus able to regulate the internal body temperature, really was a big deal for evolution!

Prior to the appearance of the birds, all of the other vertebrates (such as fish, amphibians, and reptiles), were classified as *heterothermic* (**HET**-ur-uh-**therm**-ik) or *poikilothermic* (**poy**-kih-low-**THER**-mik). This literally means that they had a "changeable" (*poikilo*) or "differing" (*hetero*) body "heat" (*therm*). That is, before the birds, the existing vertebrates had their body temperatures closely tied to the changes occurring in their surrounding external environment. Since they could not self-regulate their own body temperature, they had to resort to various behaviors (such as seen in a turtle or lizard sunning and warming itself on a rock).

Because of this body temperature limitation, reptiles and amphibians, for instance, are only able to live in fairly warm habitats. After all, you don't see many snakes or frogs near the Arctic Circle! With the appearance of birds, however, all that soon changed! Birds now dwell in nearly every corner of the world, all the while successfully being homeothermic and regulating their own body temperatures.

Probably one of the first vertebrates to make this dramatic transition from heterothermic (poikilothermic) reptile to homeothermic bird was a strange creature called *Archaeopteryx* (**ar**-kee-**AHP**-ter-iks). Its odd name, Archaeopteryx, means "ancient" (*archeo*) "bird or wing" (*pteryx*). A look at Figure 12.7 clearly shows the reasons. Archaeopteryx had such reptile-like

4, Web

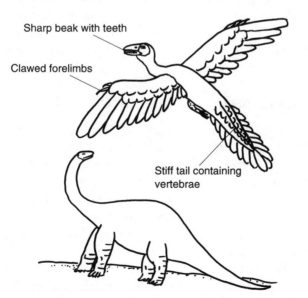

Sharp beak with teeth

Clawed forelimbs

Stiff tail containing vertebrae

Fig. 12.7 Archaeopteryx: The ancient bird-reptile.

characteristics as sharp teeth in its mouth, forelimbs with claws, and a long tail containing vertebrae. Yet, it still was part bird, because it sported a beak, and showed two wings with feathers. Archaeopteryx lived approximately 150 million years ago, and probably flew over the heads of some dinosaurs!

CLASS MAMMALIA: BACKBONED ANIMALS WITH ''BREASTS''

True to their name, mammals are vertebrates that have *mammary* (**MAM**-ah-ree) *glands* within their ''breasts'' (*mamm*). We, as human beings, of course, share such *mammalian* (mah-**MAY**-lee-un) characteristics as the nursing of the young with milk secreted by the mammary glands of the mother's breasts. Mammals also have bodies that are covered with hair, rather than with feathers (although hair, like feathers, is rich in the waterproof keratin proteins).

Mammals are like birds in that they are endothermic (as was noted back in Chapter 3). This means that their body ''heat'' (*therm*) is regulated from ''within'' (*endo*-). (Thus, endothermic is a close synonym or relative of the word homeothermic.)

5, Web

Another similarity between birds and mammals is their ancestry. It is quite likely that both classes of ''higher'' vertebrates evolved from the same ''lower'' class – the reptiles. While Archaeopteryx and other early birds flew over the heads of the dinosaurs, the first mammals were probably small and rodent-like. They probably hid in the forest and ate insects, or scurried to hide in holes below the huge, lumbering feet of the dinosaurs!

The three major groups of mammals

After an ancient comet or asteroid struck the Earth and created a massive cloud of dust and debris, the lack of sunlight and progressive cooling and drying of the climate probably contributed to the extinction of the dinosaurs. With these ''Terrible Lizards'' or Kings of the Reptiles gone from the scene, the small, rodent-like mammals had a giant *niche* (nich) or empty ''nest'' of environmental roles and functions ready to be filled.

During the Cenozoic Era or ''Age of Mammals'' (Chapter 3), stretching from about 65 million years ago up to the present, three major groups of mammals have extensively evolved. These are called the *monotremes* (**MAHN**-uh-treems), *marsupials* (mar-**SOO**-pee-als), and the *placentals* (plah-**SEN**-tals) (see Figure 12.8).

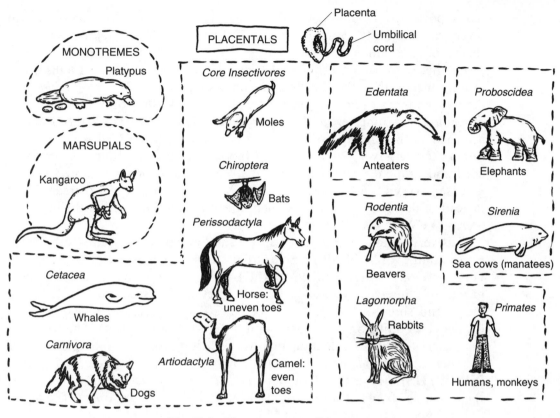

Fig. 12.8 The great story of the mammals.

The monotremes are literally mammals with a "single" (*mono-*) "hole" (*trem*)! This odd name reflects the fact that monotremes have just a single, common opening for both their urinary and reproductive tracts. So, besides excreting urine, the monotremes also lay their eggs through this single outlet. As the only group of egg-layers, the monotremes are considered the most primitive type of mammals (closest to the reptiles). Today, they are mainly represented by the *platypus* (**PLAT**-uh-pus) and the *spiny anteaters*. The platypus is named for its "flatfootedness," but an even more obvious characteristic is its prominent duckbill. The duckbilled platypus lives in Australia and New Guinea, mainly feeding upon insects.

Marsupials are mammals with "little pouches" (*marsupi*). In the wilds of North America, the most familiar marsupial is probably the *opossum*. But as captives in zoos, we usually associate pouches with the kangaroo. The *koala* bear of Australia is another famous marsupial. Rather than hatching from eggs (like the monotremes), the young marsupials are *viviparous* (vy-**VIP**-ur-us) – "born" (*par*) "alive" (*vivi*). But, since they are born almost naked and at a very early stage, the young marsupials must finish their development within their mother's pouch.

Finally, most modern mammals are placentals. In everyday English, this exactly means that their young are nourished by a "flat cake" (*placent*), actually an internal organ called the *placenta* (plah-**SEN**-tuh). During development, the youngster in its embryo and later *fetal* (**FEE**-tal) stage is attached to the placenta organ by an *umbilical* (um-**BILL**-uh-kul) or "pertaining to the navel" *cord*. The fetal blood circulates through the umbilical cord and into the placenta, where it releases waste products and picks up glucose, oxygen, and other nutrients from the mother's bloodstream. And instead of developing within a pouch, the fetus develops within the mothers' *uterus* (**YEW**-ter-us); that is, her "womb" (*uter*).

This very large group of modern placental mammals is usually subdivided into a number of highly distinctive Orders. The *core insectivores* (in-**SEK**-tuh-vors) or *Order Insectivora* (**in**-sek-**TIH**-vor-ah), for instance, are the basic group of "insect devourers" (*vores*). The core insectivores include the moles and shrews. Also eating insects are the members of *Class Edentata* (**ee**-den-**TAY**-tah), or mammals "without" (*e-*) "teeth" (*dent*). The toothless (or nearly toothless) Edentata involve such long-tongued mammals as the *anteaters* and *armadilloes*, and such "slow"-moving plant-eaters as the *sloths* (**SLAWTHS**). Members of the *Order Chiroptera* (keye-**RAHP**-ter-ah) have their "hands" (*chir*) or forelimbs modified to form "wings" (*pter*). The night-flying, insect-eating bats obviously belong to this Order.

Order Rodentia (row-**DEN**-shah), quite differently, embraces the "gnawers" with big, chisel-like teeth. Rats, mice, hamsters, beavers, squirrels, and porcupines are all *rodents*. The *Order Lagomorpha* (**lag**-uh-**MOR**-fuh) encompasses the huge group of long-eared, chisel-teethed, "hare" (*lago*) "shaped" (*morph*) mammals, such as the modern rabbits and hares.

When looking at toes rather than teeth or ears, we consider the *Order Artiodactyla* (**ar**-tee-oh-**DAK**-tuh-luh) and the *Order Perissodactyla* (puh-**ris**-uh-**DAK**-tuh-luh). The Artiodactyla are four-footed, hoofed mammals with an "even-number" (*artio*) of "toes" (*dactyl*), usually either two or four per foot. This even-numbered toe group claims the camels, deer, pigs, cattle, and sheep. The Perissodactyla, in marked contrast, are those hoofed mammals having an "uneven" (*perisso*) number of "toes" (*dactyl*) on each foot. Horses and rhinoceroses are familiar members of this uneven-toed Order.

An elephant, of course, has a very long nose or *proboscis* (proh-**BAHS**-is) – an "elephant's trunk or means of providing food." Elephants, therefore, belong to the *Order Proboscidea* (**pro**-bah-**SID**-ee-uh), because of their long, tube-like proboscis (trunk) they use for gathering leaves and other food.

Fairly closely related to the trunked Proboscidea are the members of the *Order Sirenia* (**sigh-REE**-nee-ah) – consisting of the *sirens* (**SIGH**-rens) or "sea nymphs." According to ancient Greek and Latin mythology, the sirens or sea nymphs (later called mermaids) were half woman and half fish creatures of the sea who lured sailors to their deaths on rocky shores by seducing them with their bewitching singing! In modern times, the seductive "sea nymphs" are now known as the *manatees* (**man**-uh-**TEES**). But one look at the manatee, also called the sea cow, will reveal a big, chubby, torpedo-shaped body with fin-like forelimbs, and no hindlimbs. The manatees are sea cows in much the same sense as the land-dwelling cows busily munching in farmer's fields; they are *aquatic* (uh-**KWAT**-ik) or "water (*aqua*)-dwelling" creatures that are *herbivores* (**HER**-buh-vors), "devourers" (*vor*) of "herbs or plants" (*herbi*).

Real cows, of course, have milk-giving udders. And the plant-eating manatee herbivores are named for their "female breasts" (*manati*). According to the *World Book Dictionary*, "The manatee . . . by its quiet breathing and gentle breasts probably originated the haunting mermaid legends."

Starkly opposite to the herbivores in their feeding habits are the members of the *Order Carnivora* (car-**NIV**-er-uh). The *carnivores* (**CAR**-nih-vors) are the "flesh-eaters." These flesh-eaters have specialization of their teeth for gripping and tearing. The large, sharp *canines* (**KAY**-nines) of dogs, wolves, and foxes, for example, help tear meat apart when eating it.

The *Order Cetacea* (suh-**TAY**-shuh) consists of the "large sea animals," such as the dolphins, porpoises, and whales. These *cetaceans* (suh-**TAY**-shuns) are actually a group of marine mammals that have sleek, fish-like bodies. They sport paddle-shaped forelimbs, but no hindlimbs. Even though they may casually resemble fish, they are homeothermic (endothermic) like other mammals, and nurse their young with milk.

Finally, we come to our own group, the Order Primates. The primates were briefly introduced back in Chapter 3. Recall that they are considered "of first rank or importance" (*primat*). Why? Naturally, because they include us, *Homo sapiens*! We members of *Homo sapiens* are classified as *omnivores* (**AHM**-nuh-vors). Translated into Common English, that means we are "greedy eaters" (*vores*) of almost "everything" (*omni*)! Hence, in contrast to those mammalian Orders that only include herbivores, and those that only contain carnivores, members of *Homo sapiens* have the great advantage of being very flexible in their diet. Perhaps this flexibility in diet, combined with great *manual* (**MAN**-yew-al) *dexterity* (deck-**STAIR**-uh-tee) – skillful use of the "right" (*dextr*) and left "hands" (*manu*) – has allowed our species to use its superior intelligence to prevail. As a result, members of *Homo sapiens* have put all of the other mammals into zoos (not allowing all of the other mammals to put us into the zoos)!

Figure 12.8 provides an extensive summary diagram of the three major groups and all of the important Orders of the mammals. [**Study suggestion:** After each technical term and mini-picture in the diagram, see if you can remember the literal English translation. These translations help tell "The Great Story of the Mammals." Note the groups of mammals that are most closely related, indicated by their encircling within a dotted line. Look back over the last few pages of the book to check your answers.]

Quiz

Refer to the text in this chapter if necessary. A good score is at least 8 correct answers out of these 10 questions. The answers are found in the back of this book.

1. Every chordate has:
    (a) No backbones, but a large post-anal tail
    (b) Three or more legs
    (c) A hollow nerve cord, but without any muscular tail
    (d) A notochord plus gill slits

2. Neither sea squirts nor lancelets are considered "true" vertebrates, since:
    (a) They both have closed circulatory systems
    (b) Each has a notochord during some time of its life, but never a backbone
    (c) Their backs are open and filled with seawater
    (d) Biologists are not quite sure of their evolutionary origins

3. Fish-like vertebrates "without jaws":
    (a) Class Aves
    (b) Class Agnatha
    (c) Class Amphibia
    (d) Class Placodermi

4. The Placodermi represent:
    (a) An important group of living amphibians
    (b) A long-extinct class of armored fishes
    (c) Essentially the same organisms as do modern lampreys
    (d) Close relatives of the monotremes

5. Members of the Class Chondrichthyes basically differ from those of the
    Class Osteichthyes in that:
    (a) Osteichthyes includes salamanders
    (b) Chondrichthyes involves the bony fishes
    (c) Osteichthyes consists of fish, only
    (d) Chondrichthyes is composed of fish with skeletons of cartilage

6. An amnion:
    (a) Takes its name from its remarkable resemblance to a "little
         peanut"
    (b) Chiefly consists of a fluid-filled sac encasing the embryos of higher
         vertebrates
    (c) Principally entails a naked embryo surrounded by a hard shell
    (d) Represents a stage of reptiles with gills

7. Archaeopteryx is important for biological study because:
    (a) This ancient vetebrate may have been a bridge between reptiles and
         modern birds
    (b) It probably did not appear until after all the dinosaurs had gone
    (c) There are many marsupials that resemble this creature
    (d) It likely gave birth to live young

8. To say that modern birds are homeothermic is basically to claim that
    they:
    (a) Cannot easily control their own internal body temperature
    (b) Do not have the capability to adapt to extremely cold
         environments
    (c) Can keep their body temperatures in a relatively stable range
    (d) Have lost their appetite for insects

9. The Class Mammalia get their name from what obvious characteristic?
    (a) Hairy bodies
    (b) Breasts containing milk-secreting glands
    (c) Thick craniums holding large brains
    (d) Viviparous birthing without eggs

10. The three major groups of mammals:
     (a) Aves, Reptilia, Amphibia
     (b) Koalas, opossums, and platypuses
     (c) Monotremes, marsupials, and placentals

The Giraffe ORDER TABLE for Chapter 12
(Key Text Facts About Biological Order Within An Organism)

1. _____
2. _____
3. _____
4. _____

The Spider Web ORDER TABLE for Chapter 12
(Key Text Facts About Biological Order Beyond The Individual Organism)

1. _____
2. _____
3. _____
4. _____
5. _____

The Broken Spider Web DISORDER TABLE for Chapter 12
(Key Text Facts About Biological Disorder Beyond The Individual Organism)

1. _____

Test: Part three

DO NOT REFER TO THE TEXT WHEN TAKING THIS TEST. A good score is at least 18 (out of 25 questions) correct. Answers are in the back of the book. It's best to have a friend check your score the first time, so you won't memorize the answers if you want to take the test again.

1. Biological discipline which defines and classifies different groups of organisms:
   (a) Ecology
   (b) Anatomy
   (c) Taxonomy
   (d) Physiology
   (e) Bacteriology

2. The bacteria and all other types of prokaryotes are members of this kingdom:
   (a) Bacteriae
   (b) Monera
   (c) Plantae
   (d) Animalia
   (e) Fungi

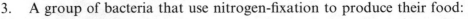
3. A group of bacteria that use nitrogen-fixation to produce their food:
    (a) Photoautotrophs
    (b) Heterotrophs
    (c) Chemoautotrophs
    (d) Gram-positive
    (e) Gram-negative

4. Comes from the Latin for "poison":
    (a) Bacteremia
    (b) Septicemia
    (c) Antibiotics
    (d) Virus
    (e) Coccus

5. Extensively use pseudopodia to help them eat and move:
    (a) Euglena
    (b) Bacteria
    (c) Amoebas
    (d) Mushrooms
    (e) Plankton

6. Type of fungi that get much of their nutrition from rotting plants and leaves:
    (a) Parasitic
    (b) Pathogenic
    (c) Saprophytic
    (d) Gram-positive
    (e) Photosynthetic

7. Every time you see the suffix, -mycetes, you know that a _____ is somehow involved:
    (a) Fungus
    (b) Plant
    (c) Virus
    (d) Bacterium
    (e) Amoeba

8. Alcoholic fermentation is primarily carried out by:
    (a) Morel mushrooms
    (b) Anaerobic yeast cells
    (c) Hyphae
    (d) Oxygen-utilizing club fungi
    (e) Fertile basidia

9.   Mixtures of fungi with algae or bluish-green bacteria in symbiosis:
     (a) Chloroplasts
     (b) Lichens
     (c) Cellular slime molds
     (d) Flagella
     (e) Mosses

10. Scattered white patches appearing on a person's mucous membranes
     may well signal the presence of:
     (a) Albinism
     (b) Enhanced immune system activity
     (c) Candidiasis
     (d) Amebiasis
     (e) Deadly bark-rot syndrome

11. A partial, undeveloped plant contained within a seed:
     (a) Zygote
     (b) Ovum
     (c) Sperm
     (d) Embryo
     (e) Fetus

12. Mosses and other moss-like plants:
     (a) Nonvascular plants
     (b) Tracheophytes
     (c) Vascular plants
     (d) Oviducts
     (e) Lichens

13. Sphagnum is also known as:
     (a) Peat moss
     (b) Water chestnut
     (c) Plantain
     (d) Leafy veins
     (e) Chlorophyll

14. Lacking a vascular (vessel-bearing) system, they must grow upon
     shallow pools of water:
     (a) Conifers
     (b) Gymnosperms
     (c) Mosses and liverworts
     (d) Lichens and corals
     (e) Club fungi and sac fungi

15. Ferns:
    (a) Have split leaves or fronds
    (b) Show broad, flat leaves
    (c) Cannot reproduce naturally
    (d) Grow only in cold, dry climates
    (e) Did not exist during the Carboniferous Period

16. An entire pine tree is technically called a:
    (a) Nonvascular tracheophyte
    (b) Mature sporophyte
    (c) Partial basidium
    (d) Amoeboid center
    (e) Oogonium

17. "Gnetophytes, like all _____, reproduce by means of spores and cones":
    (a) Gymnosperms
    (b) Angiosperms
    (c) Spermatozoa
    (d) Oocytes
    (e) Zygotes

18. The female reproductive organ of a flower:
    (a) Anthers
    (b) Filaments
    (c) Stamen
    (d) Corolla
    (e) Carpel

19. Plucking one petal from an otherwise perfect round flower would result in:
    (a) Creation of symmetry
    (b) Enhanced Biological Order
    (c) "The process of measuring together"
    (d) Introduction of bilateral symmetry
    (e) Breaking of radial symmetry

20. After a zygote undergoes several cleavages, the eight-cell stage is followed by:
    (a) A blastula, then a gastrula
    (b) An archenteron, then a blastula
    (c) A gastrula, then a blastula
    (d) Endosymbiosis
    (e) Self-pollination

21. Bilateral invertebrates with no main body cavity:
     (a) Annelids and other types of coelomates
     (b) Primates and many other mammals
     (c) Planaria and other types of acoelomates
     (d) Nearly all types of autotrophs
     (e) Mollusks

22. A huge phylum of invertebrates with segmented bodies and "jointed feet":
     (a) Humans and other vertebrates
     (b) Fungi
     (c) Arthropods
     (d) Plantae
     (e) Mycetes

23. Comprise the hugest number of known species of any animal:
     (a) Millipedes
     (b) Echinoderms
     (c) Centipedes
     (d) Marsupials
     (e) Insects

24. Evolutionary "bridging species" between most invertebrates and the true vertebrates:
     (a) Archaeopteryx and other ancient "bird-reptiles"
     (b) Monarch butterflies and pupating moths
     (c) Sea squirts and lancelets
     (d) Sharks and other cartilaginous fishes
     (e) Snakes and turtles

25. The vertebrates all have a (an) _____ _____ system:
     (a) Cartilaginous osseous
     (b) Open circulatory
     (c) Nonfunctional nervous
     (d) Open-topped vertebral
     (e) Closed circulatory

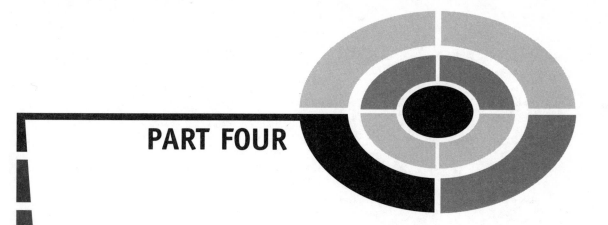

PART FOUR

Anatomy and Physiology of Animals

13

Skins and Skeletons

Having introduced and characterized the mammals, it is now appropriate for us to further examine the meaning of the following claim: "Oh, those pesky animals! They're just a bag of bones!" Yes, this is true! The "bag" is their surrounding skin, while the "bones" are the major organs making up their skeletons.

Yet, there is also another reason why we consider both the skin and the skeleton together, here in Chapter 13. Recall (Chapter 10) that in all animals, except for sponges, there are two or more germ layers within the embryo. These germ layers are technically called *derms* or "skins," since they consist of thin groups of primitive cells that progressively *differentiate* (**dih**-fer-**EN**-she-ate) or "become more different," as the embryo undergoes its development. In the hollow gastrula and later stages of embryonic development, there are three derms or skins. Remember that Chapter 10 identified these as the endoderm (inner skin), mesoderm (middle skin), and ectoderm (outer skin).

The ectoderm eventually forms much of the skin, hair, nails, the enamel of the teeth, and the most important portions of the nervous system. The mesoderm, in contrast, ultimately gives rise to the bones, muscles, much of the heart and circulatory system, and the connective tissues. We consider skin and skeleton together in the same chapter because they arise from two adjacent germ layers within the embryo.

1, Order

217

Skins and Skeletons: Tough Reminders of a Former Life

But the final reason we lump them together is due to the answer to the following question: "When you are out walking in the woods or fields, and you come upon the remains of a long-dead animal, what part of the body is usually left for you to identify?" Certainly, the two most prominent (and perhaps toughest) parts of the dead animal are its covering of skin (with maybe attached fur or feathers) on the outside, and its bony skeleton lying within. Indeed, paleontologists (scientists who study ancient life) most frequently use skeletal remains – teeth, and (if they're very lucky) fragments of preserved hide or skin – to help them identify and classify long-dead or extinct animals. Thus, we study skin and bones together, because that's all what is usually left after an animal dies!

"Ain't We Just Peachy?": The Skin as Our Integument

"How is the human body like a peach?" In reply to this odd question, you might comically retort, "Well, we're just a bunch of fruits!" But more seriously, both a peach and the human body are surrounded by an *integument* (in-**TEG**-you-ment) – a thin "covering." And lying deep to this integument is a much thicker layer of flesh. Hence, it is the integument or "skin" (derm) that encloses and protects the flesh.

And the skin or integument of both the peach and the human body has an outermost layer, the epidermis. You might remember from our study of plants (Chapter 9) that the epidermis is literally "(something) present upon the skin," and that in ferns and seedbearing plants (such as peach trees), the epidermis is an extremely thin layer of outer protective cells. But from here, the similarity between peach integument and human integument begins to fade (Figure 13.1).

KERATINIZED EPITHELIAL STRATA

One key difference between the epidermis of humans and most other animals, from those of plants, is the presence of keratin. This "horn substance"

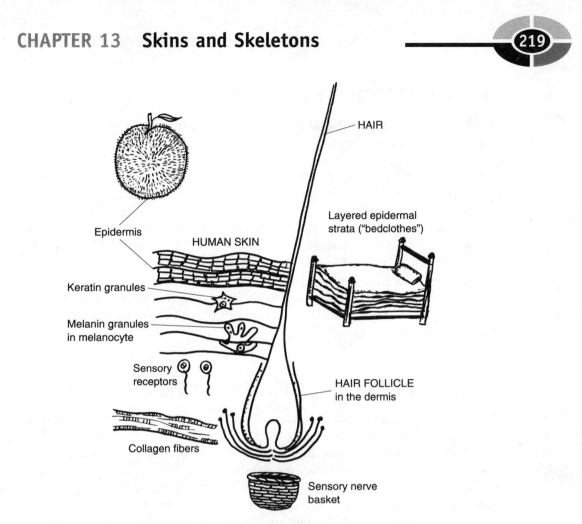

Fig. 13.1 Anatomy of the human skin: Not just a peach!

(Chapter 12) is found in the epidermis and hair and nails of humans, as well as in the claws and horns of various other animals. Since keratin is essentially waterproof, so is human skin. Our epidermis consists of a multiple series of thin, overlapping *strata* (**STRAT**-uh) of keratinized epithelial cells. When viewed under a light microscope, this highly orderly arrangement gives the epidermis the appearance of a collection of thin "layers or bed covers" (*strat*), heaped one upon the other over a bed.

2, Order

   The outer surface *stratum* (**STRAT**-um) is really not living, at all! Rather, it consists of a series of stacked *squamae* (**SKWAY**-me) – dead, keratin-stuffed "scales" (*squam*). This means that you and I face the world with dead scales (squamae) showing at our body surface! But this has the definite advantage of protecting us from dirt, fungi, and bacteria, which cannot easily penetrate the multiple layers of lifeless squamae.

SKIN COLORATION BY MELANIN

In addition to containing keratin, the epithelial cells in our epidermis are also rich in *melanin* (**MEL**-uh-nin) or "black" (*melan*) "substance" (*-in*). Melanin is a brownish-black pigment produced by *melanocytes* (**MEL**-uh-nuh-**sights**). These "black cells" are large, octopus-shaped cells with several long arms of cytoplasm. After they produce the melanin granules, the melanocytes appear to penetrate the membranes of adjacent cells, and inject some of their melanin granules into them. This results in a darkening of the epidermis. Melanin's chief function is absorption of *ultraviolet* (**ul**-truh-**VEYE**-uh-lit) *rays* that strike the surface of the skin. These ultraviolet (UV) rays are invisible rays whose wavelengths lie "beyond" (*ultra-*) those of X-rays, but below those of visible violet light. The benefits are dual: reduction in the risk of suffering skin cancer (due to mutation of skin cell DNA by UV light), and reduction in skin wrinkling.

THE DERMIS AS OUR TOUGH MAIN "SKIN"

3, Order

True to its name, the epidermis literally lies "upon" (*epi-*) the *dermis* (**DER**-mis). The dermis is the main, fiber-rich, connective tissue portion of the skin. The dermis is especially rich in a dense network of *collagen* (**CALL**-uh-jen) *fibers*. The word collagen translates to mean "glue" (*coll*) "producer" (*-gen*). Collagen fibers are thick, tough, unbranched fibers that have a high *tensile* (**TEN**-sil) *strength*; that is, they have a great ability to resist pulling or "tension" (*tens*) forces. [**Study suggestion:** Place two fingers on the skin of your forearm. Now, gently try to stretch your skin between your fingertips. Feel the resistance to stretching, the tensile strength, that is being exerted? This glue-like function is mainly the result of the thousands of collagen fibers running like a tough, woven basket throughout your dermis.]

One important type of structure found within the dermis is the *hair follicle* (**FAHL**-uh-kul). As evident from Figure 13.1, the hair follicle is a "little bag" lined by a membrane, and containing a hair. The base of the hair follicle lies in the dermis. The hair, itself, is basically a flexible rod of tightly packed, keratin-stuffed squamae. Arranged around the base of the hair follicle is a *sensory nerve basket*. [**Study suggestion:** Without touching your skin, gently stroke the hairs on your forearm. What do you feel – a tickling, tingling sensation? This reveals the main function of hairs: touch sensations.]

There are many other types of *sensory receptors* located within the dermis. Besides receptors for the sense of touch, there are those for pressure, pain, cold, heat, and vibration.

One of the most critical functions of the dermis is its role in *thermoregula-tion* (**THER**-moh-reg-you-**LAY**-shun). By thermoregulation, we mean the "regulation" or control of "heat" (*therm*): specifically, the control of internal body temperature. Way back in Chapter 1, we talked about the homeostasis or relative constancy of oral body temperature, measured in units of degrees Fahrenheit. This homeostasis (relative constancy) of oral body temperature is another term for thermoregulation. It was represented symbolically by an S-shaped curve shown back in Figure 1.2 (A):

4, Order

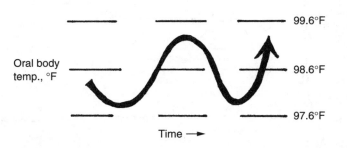

Figure 13.2 reveals what happens when the oral body temperature of a human or other homeothermic animal rises towards the upper limit of its normal range. (This temperature boost often occurs during heavy exercising.) In the dermis, two critical events kick in. The *sweat glands* increase their secretion of sweat into the *sweat ducts*, which then moves up and out onto the surface of the skin through *sweat pores*. The excess body heat essentially boils the watery sweat from the skin surface, causing it to evaporate into the air. This net heat loss helps to lower the oral body temperature.

A second chain of physiological events involve *vasodilation* (**vase**-oh-die-**LAY**-shun) – the "process of" (-*tion*) blood "vessel" (*vas*) "widening" (*dilat*). As the body gets hotter, the blood vessels in the dermis *vasodilate* (**vase**-oh-**DIE**-late), becoming wider. This allows more hot blood to circulate from the deep core of the body, and flow more freely into the vessels of the skin. Much more heat is then lost by *radiation*, the movement of heat waves or rays from the hot blood in the skin out into the cooler air surrounding the body.

By both of these means combined (increased evaporation from sweat + increased heat loss by radiation from the blood), oral body temperature is eventually brought back down to its average, long-term level (about 98.6 degrees Fahrenheit in most human beings). And thermoregulation, or home-ostasis of oral body temperature, is thereby achieved.

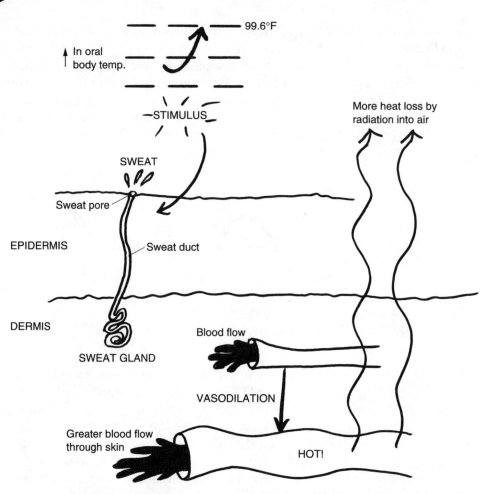

Fig. 13.2   Thermoregulation: Keeping our body heat within range.

Hyperthermia or Hypothermia: Body Heat Out of Control!

The high degree of Biological Order of body temperature shown within Figure 13.2 could formally be called a state of *normothermia* (**nor**-moh-**THER**-me-uh). Literally, normothermia is "a condition of" (*-ia*) "normal" (*normo-*) body "heat" (*therm*). Because thermoregulation is proceeding so successfully, oral body temperature does go up and down over time, yet it still remains between a low of about 97.6 degrees F and a high of approximately 99.6 degrees F. Thus, there is a homeostasis of oral body temperature, such that it remains within its "normal" (*normo-*) range over time. The person remains clinically healthy with regards to body temperature.

Now consider, in marked contrast, the abnormal conditions illustrated within Figure 13.3. What happens when oral body temperature rises significantly above the upper normal limit of 99.6 degrees F? As displayed in Figure 13.3 (A), the resulting state of Biological Disorder is technically called *hyperthermia* (**high**-per-**THER**-me-uh). This is "a condition of excessive or above normal" (*hyper-*) body heat or temperature. In hyperthermia, the patient is essentially suffering from a fever. It may be accompanied by such other familiar symptoms as chills and muscular aches and pains. If breaking of the normally S-shaped pattern of body temperature homeostasis is severe enough, then the temperature keeps rising past 108 degrees F, and the person may go into a coma and die!

Certainly, then, hyperthermia is to be avoided! "Okay, so let's drive the oral body temperature way below its lower normal limit of about 97.6

1, Disorder

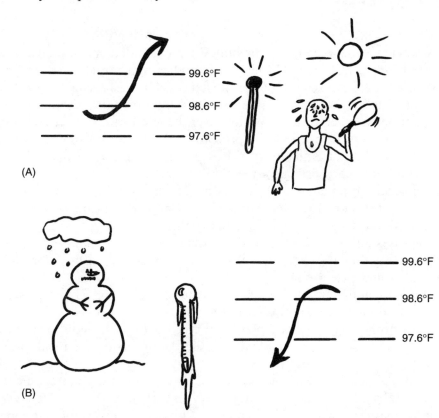

(A)

(B)

Fig. 13.3 Hyperthermia (A) and hypothermia (B): Too much, or too little, of a good thing.

degrees F!'' you might helpfully suggest at this point. ''In this way, we won't have to worry so much about hyperthermia!''

But, thinking carefully, are we still not breaking the S-shaped pattern of thermoregulation or homeostasis? Here, however, the breaking of pattern or introduction of Biological Disorder is at the low end, rather than the high end, of the body temperature range. When oral body temperature plunges significantly below 97.6 degrees F, then a state of *hypothermia* (**high**-poh-**THER**-me-uh) is said to exist. Hypothermia is a ''deficient or below normal'' (*hypo-*) ''condition of'' body ''heat.'' Since it represents temperature going down to an excessively low level, hypothermia (like hyperthermia), if great enough, can result in coma and death.

2, Disorder

The Human Endoskeleton: Our Hard ''Dried Body'' Lying within

1, Web

If we humans are just a ''bag of bones,'' then, having discussed the ''bag'' (skin), we now need to consider the bones! Recollect that crabs, lobsters, and other arthropods (Chapter 11) have soft bodies that are covered and protected by a tough exoskeleton of the substance, chitin. The skeleton or ''hard dried body'' is on the outside, like a coat of mail or battle armor. Human beings and most other vertebrates, however, have taken the exactly opposite approach to these arthropods in protecting their delicate internal organs from physical trauma and injury. They possess an *endoskeleton* (**EN**-doh-**skel**-uh-tun), or ''hard dried body within'' (*endo-*).

A big peach, in a plant-sense, has its own form of endoskeleton – a rock-hard inner *pit* (Figure 13.4). As explained in Chapter 9, a ripened fruit, such as a nice round, pinkish peach, is actually the fleshy wall of a plant ovary, which encases and protects the seeds holding the delicate plant embryos. In peaches, the stony pits are the seeds that contain the early embryos for germinating new peach trees.

In humans and our animal relatives, the endoskeleton is comprised of many individual bone organs (and the joints made between them). But instead of protecting a delicate plant embryo, bones in the skull or cranium protect the soft eyes and brain. And bones in the thorax or chest wall protect the relatively soft heart and lungs.

In both the peach and the human body, a very thin, tough integument or skin ensheaths a large quantity of relatively soft flesh. And lying deep within this fleshy mass is a protective peach pit or (in the case of vertebrate animals) a protective endoskeleton.

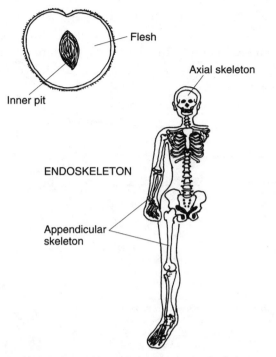

Fig. 13.4 Peaches, pits, and the human endoskeleton.

Humans, of course, have a much more complex body plan, compared to peaches! Our skeleton is subdivided into two main portions. An axial skeleton (discussed back in Chapter 12) lies within the head, neck, and body trunk. Conversely, an *appendicular* (**ah**-pen-**DIK**-you-ler) *skeleton* lies within the body *appendages* (ah-**PEN**-dah-**jes**) or limb "attachments." The appendicular skeleton consists of the bones in the upper appendages (the shoulders, arms, wrists, and hands), as well as those in the lower appendages (the hips, legs, ankles, and feet). Taken together, the axial and appendicular skeletons make up the "hard dried body" or endoskeleton lying "within" us.

And being chordates, our vertebral columns stiffen and support us from the inside, rather than from the outside.

Anatomy of a Long Bone

To fully understand the skeleton, we must examine the anatomy of a typical *long bone*. A long bone is simply a bone that is much longer than it is wide. Consider, for instance, the "thigh" bone or *femur* (**FEE**-mur). As Figure 13.5

reveals, the femur has a main shaft, called the *diaphysis* (die-**AH**-fuh-**sis**), which is literally a "growth" (*phys*) "through" (*dia-*) the middle of the bone. And capping each end of the diaphysis (main bone shaft) is an *epiphysis* (eh-**PIH**-fih-**sis**) – a "growth" (*phys*) present "upon" (*epi-*) the shaft.

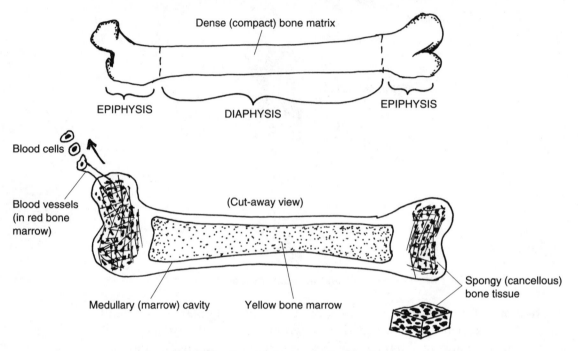

Fig. 13.5   General anatomy of a long bone.

Most of the long bone is composed of *dense or compact bone matrix* (**MAY**-tricks). This is the white, rock-hard, calcium-rich portion of the long bone. But in the "middle" (*medull*) portion of the long bone, we find the *medullary* (**MED**-you-**lair**-ee) *or marrow cavity*. This medullary (marrow) cavity, as its name indicates, contains the *yellow bone marrow*. The yellow color of this type of marrow is mainly due to the presence of *adipose* (**AH**-dih-**pohs**) or "fatty" *connective tissue*.

Finally, we see the *spongy or cancellous* (**CAN**-seh-**lus**) *bone* located within each epiphysis of the long bone. It obviously gets its spongy name from the existence of numerous holes and an extensive network of *cancelli* (can-**SEL**-eye) – "little crossbars" (*cancell*) of hard bone matrix. Much like a real sponge, therefore, spongy (cancellous) bone tissue consists of a network of cancelli or little crossbars and the many holes between them. Unlike a real sponge, however, spongy bone tissue contains *red bone marrow* within its holes. The red bone marrow is red in color mainly due to the fact that it consists of many blood vessels. These dozens of blood vessels run and branch extensively throughout the holes of the spongy bone. Their main function is that of *hematopoiesis* (**he**-muh-toh-poi-**E**-sis) – the process of "blood" (*hemat*) "formation" (*-poiesis*). Most of the blood cells (and blood cell fragments) ultimately are formed by hematopoiesis occurring within the red bone marrow. The blood cells enter the general bloodstream when they are circulated out of the long bone, through the vessels leaving the red bone marrow.

Bone Development, Bone Matrix, and Blood Calcium Homeostasis

Besides hematopoiesis and protection from physical trauma, another critical function of the endoskeleton is *blood calcium homeostasis*; that is, the maintenance of a relatively constant blood calcium ion concentration. Symbolically speaking, we use Ca^{++} to identify *blood calcium ions*, and brackets, [ ], to denote concentration. Thus, we have $[Ca^{++}]$ to indicate the blood calcium ion concentration.

5, Order

The blood calcium ion concentration, $[Ca^{++}]$, within humans, is usually measured in units of *mg/dL – milligrams* (**MIH**-lih-**grams**) of calcium ions per *deciliter* (**DEH**-sih-**lee**-ter) – of blood. A deciliter is one-"tenth" (*deci-*) of a "liter." And milligrams is a unit representing the number of "thousandths" (*milli-*) of a "gram" of some substance. Hence, blood $[Ca^{++}]$ in mg/dL denotes the number of milligrams of calcium ions present within one-tenth of a liter of blood. The normal or *reference range* for blood $[Ca^{++}]$ is from a low of about 8.5 to a high of approximately 10.6 mg Ca^{++}/dL of blood (in adults). Taking the same approach we employed for thermoregulation (Figure 13.2), we can use the S-shaped pattern, once again, to represent the homeostasis of blood calcium ion concentration, over time (see Figure 13.6). In naming this particular pattern of chemical concentration, we use the suffix, *-emia* ("blood condition of"), and the root or main idea, *calc* ("calcium"). We therefore have some form of *calcemia* (kal-**SEE**-me-uh), or "condition of calcium" (ion concentration) within the "blood." [**Study suggestion:** Using the same prefix as that employed to describe the normal-range body temperature pattern of Figure 13.2, name the blood calcium ion concentration pattern symbolized by Figure 13.6, below. When you are done building this term, check it with the correct term found in the caption for Figure 13.6.]

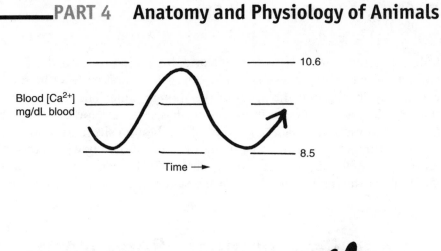

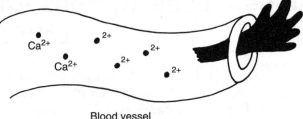

Fig. 13.6 Maintenance of normocalcemia over time.

Bone development and bone matrix

"How is normocalcemia related to bone development and bone matrix?" you might well ask at this time. The answer is provided by a close look at Figure 13.7. During the development of a long bone, such as the femur ("thigh" bone), the process essentially begins with a *cartilage model* – a miniature version of the bone that is composed of cartilage, rather than bone tissue. Being soft and rubbery, cartilage is more suitable for life within the mother's uterus (womb). As development progresses, however, blood vessels break into the cartilage model and bring *osteoblasts* (**AHS**-tee-oh-**blasts**) along with them. Osteoblasts are literally "bone" (*oste*) "formers" (-*blasts*).

The osteoblasts are large, spider-shaped cells that produce bone collagen fibers. After these tough collagen fibers are laid down, the osteoblasts then extract Ca^{++} ions, phosphorus, and other chemicals from the bloodstream. *Ossification* (**ah**-sih-fih-**KAY**-shun), the "process of bone formation," then begins. During ossification, the osteoblasts supervise the depositing of sharp, needle-shaped crystals onto the surfaces of the bone collagen fibers. These sharp crystals are composed of *calcium phosphate* (**FAHS**-fate), as well

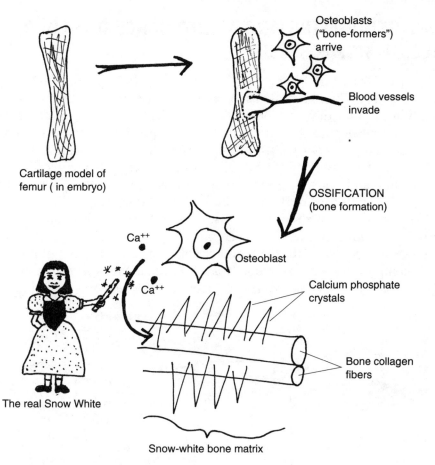

Osteoblasts
("bone-formers")
arrive

Blood vessels
invade

Cartilage model of
femur (in embryo)

OSSIFICATION
(bone formation)

Ca^{++}

Osteoblast

Ca^{++}

Calcium phosphate
crystals

Bone collagen
fibers

The real Snow White

Snow-white bone matrix

Fig. 13.7   Bone matrix appears during ossification.

as a number of other minerals. Essentially, bone matrix appears within the cartilage model of the long bone, because the newly produced bone collagen fibers become heavily coated with the calcium phosphate crystals. Being snow-white, these crystals eventually hide the underlying collagen fibers. Viewed with the naked eye, the entire bone matrix thus appears snow-white. The bone matrix is white, like cement, of course, because white is the color of the many thousands of calcium crystals covering the collagen fibers in the matrix.

By the time the child becomes an adult, her femur is mostly snow-white. You might think of all the bone matrix in the femur as essentially being a storage bank for calcium ions.

BONE REMODELING AND MAINTENANCE OF BLOOD CALCIUM HOMEOSTASIS

Just as osteoblasts are "bone-makers," cells called *osteoclasts* (**AHS**-tee-oh-**klasts**) are "bone-breakers" (-*clasts*). By bone-"breakers," of course, we don't literally mean that these osteoclast cells actually break or fracture the bone! Rather, we mean that the osteoclasts release special digestive enzymes that cause a partial *resorption* (rih-**SORP**-shun) or "drinking-in again" of the bone matrix, such that some of the calcium phosphate crystals are dissolved back into calcium and phosphate ions. Such a process often occurs whenever the blood $[Ca^{++}]$ falls toward the lower limit of its normal range (Figure 13.8, A). And as a result of this resorption (dissolving or "drinking-in again") of bone matrix, Ca^{++} ions that were formerly stored in the bone matrix, like a bank, are now released back into the blood circulation. This process, quite clearly, acts to temporarily raise the blood $[Ca^{++}]$. Because the bone gets thinner and weaker, we say it has *remodeled* – changed its thickness and strength.

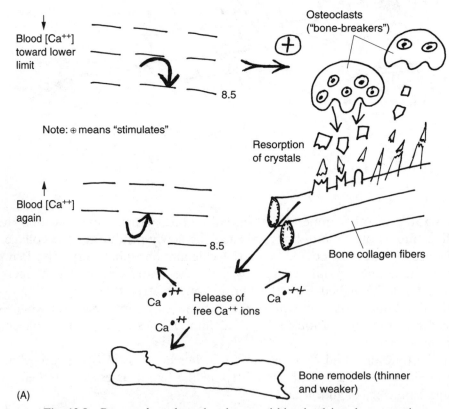

Blood $[Ca^{++}]$ toward lower limit

8.5

Note: ⊕ means "stimulates"

Osteoclasts ("bone-breakers")

Resorption of crystals

Blood $[Ca^{++}]$ again

8.5

Bone collagen fibers

Ca^{++} Release of free Ca^{++} ions Ca^{++}

Ca^{++}

Bone remodels (thinner and weaker)

(A)

Fig. 13.8 Bone-makers, bone-breakers, and blood calcium homeostasis.

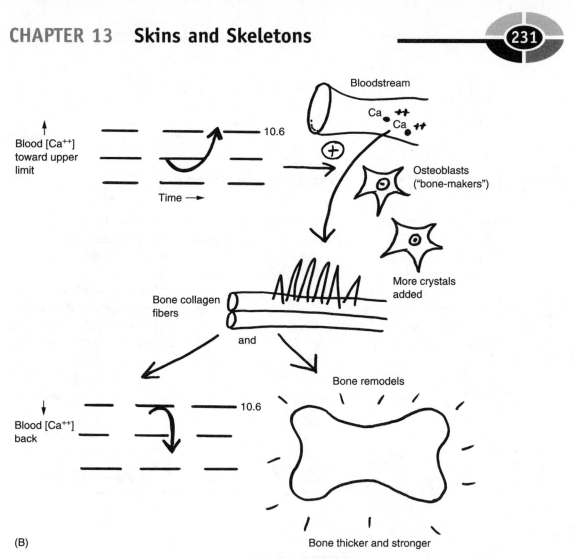

Fig. 13.8   (continued)

Conversely, whenever a person eats a calcium-rich meal, the blood [Ca^{++}] rises toward the upper limit of its normal range (Figure 13.8, B). Thousands of osteoblasts are stimulated, then more Ca^{++} ions are extracted from the bloodstream and laid down on the bone collagen fibers as calcium phosphate crystals. The bone once again remodels (changes its shape and size), but this time it goes in the opposite direction, by getting thicker and stronger. And as more and more free Ca^{++} ions are extracted from the bloodstream and put into calcium phosphate crystals, the blood [Ca^{++}] falls.

By these contrasting processes involving osteoblasts, osteoclasts, bone remodeling, and bone resorption, therefore, homeostasis of blood [Ca^{++}] is usually maintained. Maintaining a normal blood [Ca^{++}] is critical for human health. Why? One reason is that the contractions of all our body muscles (including those of the heart) require an adequate level of calcium ions within the bloodstream.

Breaking of Bone-related Patterns

We can think of the normal anatomy of a long bone (such as the femur) as an example of Biological Order at the organ level of body organization. And blood [Ca^{++}] homeostasis likewise provides a model of Biological Order at the chemical level of biological organization (Chapter 2).

What, then, about Biological Disorder for the above examples? What, specifically, could go wrong with each of them? A long bone could suffer a fracture (Figure 13.9, A), thereby breaking its normal anatomical pattern. And blood [Ca^{++}] could rise far above its normal range (Figure 13.9, B). Conversely, it could fall far below its normal range (Figure 13.9, C). [**Study suggestion:** Using the term, normocalcemia, along with an appropriate prefix, build a term that describes the blood [Ca^{++}] in Figure 13.9, (B); now build

3, Disorder

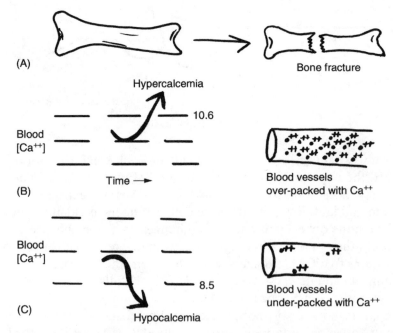

Fig. 13.9　Breaking some bone-related patterns.

another term summarizing the blood [Ca^{++}] in Figure 13.9, C. When done, check your answers with the terms given in Figure 13.9.]

Joints: A Meeting of the Bones

A junction is a meeting or joining place. (Picture a railroad junction or crossing of two intersecting train tracks.) A *joint*, therefore, is a place of meeting or joining between bones. There are three main types or categories of joints within the human body (Figure 13.10).

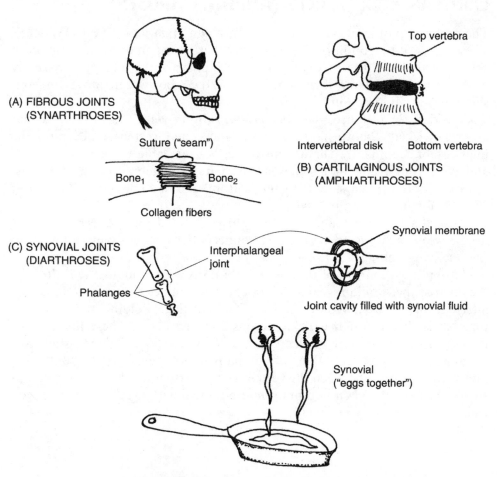

Fig. 13.10 The three main types of joints.

FIBROUS JOINTS (SYNARTHROSES)

The simplest group are the *fibrous* (**FEYE**-brus) *joints* or *synarthroses* (**sin**-ar-**THROW**-seez). As their name indicates, the adjacent bones in a fibrous joint are more-or-less strapped together by a set of collagen fibers. A good example is provided by the *sutures* (**SOO**-churs) or jagged "seams" (*sutur*) running between individual skull and facial bones.

The sutures and other types of fibrous joints are immovable. This gives them the alternate name of synarthroses – literally "conditions of" (*-oses*) "joints" (*arthr*) with the bones strapped "together" (*syn-*).

CARTILAGINOUS JOINTS (AMPHIARTHROSES)

The next group of joints are called *amphiarthroses* (**am**-fee-ar-**THROW**-seez). These are "joint conditions" (*arthroses*) permitting movement "on both sides" (*amphi-*) of the involved bones. Amphiarthroses are only partially movable joints, but (as their name states) their bones can move on both sides, and in all directions.

Consider, for instance, the *intervertebral* (**in**-ter-ver-**TEE**-bral) *joints* that are sandwiched "between" (*inter-*) the individual vertebrae. Because the intervertebral joints are slightly movable, you are able to bend your back and twist your trunk moderately as the jointed vertebrae move short distances on both of their sides (top and bottom) and in all directions. You may even be able to dance the "twist"!

An oval slab of cartilage connective tissue, called the *intervertebral disc*, lies between each two vertebrae and connects the vertebra lying above it to the one lying below.

Cutting a section out of an intervertebral disc, and looking at it through the microscope, would reveal a slab of cartilage connective tissue. (Hence the alternate name, cartilaginous joints.) The cartilage holding the vertebrae together is shot through with numerous collagen fibers. These fibers make the oval, disc-shaped slab of cartilage behave much like a stale marshmallow or tough cushion. Because there are so many intervertebral joints (containing intervertebral discs) between the vertebrae, the vertebral column has excellent shock absorption. When you jump up-and-down with excitement, then, you usually don't break your back!

SYNOVIAL JOINTS (DIARTHROSES)

The third main category of joints are the *diarthroses* (**die**-ar-**THROW**-seez) or *synovial* (sin-**OH**-vee-al) *joints*. The word, *diarthrosis* (**die**-ar-**THROW**-sis), indicates a "double" (*di-*) "joint" (*arthr*) "condition" (*-osis*). An individual is said to be "double-jointed" when the *interphalangeal* (**in**-ter-fah-lan-**GEEL**) *joints* "between" (*inter-*) his "finger or toe bones" (*phalange*) have an unusually high degree of mobility. Hence, the diarthroses are the freely movable joints, with bones so movable that many seem to be double-jointed!

Finally, the word, synovial, "pertains to" (*-al*) "eggs" (*ovi*) "together" (*syn-*). The amusing thinking behind this name reflects the appearance of the *synovial fluid*. This fluid is secreted by the *synovial membrane* lining the hollow *joint cavity*. The synovial fluid is clear, thick, and slimy, making it look like the raw white portion of many eggs poured into a frying pan together. Due to its slippery nature, the synovial fluid significantly reduces bone friction and wear while the body carries out most of its major movements.

Quiz

Refer to the text in this chapter if necessary. A good score is at least 8 correct answers out of these 10 questions. The answers are found in the back of this book.

1. In general, differentiation is:
   (a) The process whereby the individual germ layers in the embryo eventually become highly specialized
   (b) A gradual reduction in complexity of body tissues with increasing age
   (c) Any biological means by which organs become more similar to each other
   (d) A specific type of diffusion of molecules

2. One major reason for studying the skin and skeleton together within the same chapter:
   (a) The skin and bones are both primarily composed of calcium crystals
   (b) Both organ systems remain to help identify an animal after it dies
   (c) Because all the other organ systems fit together as one
   (d) Purely a matter of random guesswork!

3. Unlike plants, the epidermis of humans and most other animals:
    (a) Is rich in keratin proteins
    (b) Has adequate reserves of chlorophyll
    (c) Participates in photosynthesis
    (d) Is located deep (internal) to the actual body surface

4. Dead, waterproof "scales" lying upon the skin surface:
    (a) Squamae
    (b) Strata
    (c) Melanocytes
    (d) Granules

5. Collagen fibers are very tough, due to their high:
    (a) Elasticity
    (b) Tension
    (c) Rate of excretion
    (d) Tensile strength

6. The skin primarily achieves its thermoregulation through the processes
    of _____ and _____:
    (a) Sweat evaporation and vasodilation
    (b) Radiation and vasoconstriction
    (c) Sweat evaporation and cell loss
    (d) Radiation and sensory reception

7. Unlike crabs and other arthropods, human beings have:
    (a) An exoskeleton
    (b) Jointed appendages
    (c) An endoskeleton
    (d) Chitin in their skin

8. The main shaft of a long bone:
    (a) Diaphysis
    (b) Spongy spone
    (c) Epiphysis
    (d) Marrow

9. Hematopoiesis is a critical function of:
    (a) Red bone marrow
    (b) Cancelli
    (c) Yellow bone marrow
    (d) Adipose connective tissue

10. The sharp crystals covering bone collagen fibers are largely composed of:
    (a) Calcium phosphate
    (b) Magnesium sulfide
    (c) Chlorophyll pigment
    (d) Melanin

The Giraffe ORDER TABLE for Chapter 13
   (Key Text Facts About Biological Order Within An Organism)

1. _____
2. _____
3. _____
4. _____
5. _____

The Dead Giraffe DISORDER TABLE for Chapter 13
   (Key Text Facts About Biological Disorder Within An Organism)

1. _____
2. _____
3. _____

The Spider Web ORDER TABLE for Chapter 13
(Key Text Facts About Biological Order Beyond the Individual Organism)

1. _____

The Neuromuscular (Nerve–Muscle) Connection

Chapter 13 discussed the structure and function of the skin and skeleton. Chapter 14 considers the *neuromuscular* (**nur**-oh-**MUS**-kyoo-lar) or "nerve" (*neur*) and "muscle" (*muscul*) connection. Quite fascinating is that the word *muscular* (**MUS**-kyoo-lar) actually translates to mean "pertaining to a little mouse (*muscul*)"!

Literally speaking, then, the *muscular system* of humans is an organ system composed of over 600 individual "little mice" – the *skeletal muscle organs*. In reality, the skeletal muscle organs are a large group of individual skeletal muscles that get their name from the fact that they are attached to the bones of the skeleton (Figure 14.1). [**Study suggestion:** Flex or bend your forearm. Now, extend or straighten it. Observe how your *biceps* (**BUY**-seps) *brachii* (**BRAY**-kee-eye) muscle bulges, then lengthens, deep to the skin in your upper arm. Try to duplicate the imagination of the early anatomists, by

1, Order

speculating why they called muscles like the biceps brachii the "little mice." Check your thinking with the diagram in Figure 14.1, below.]

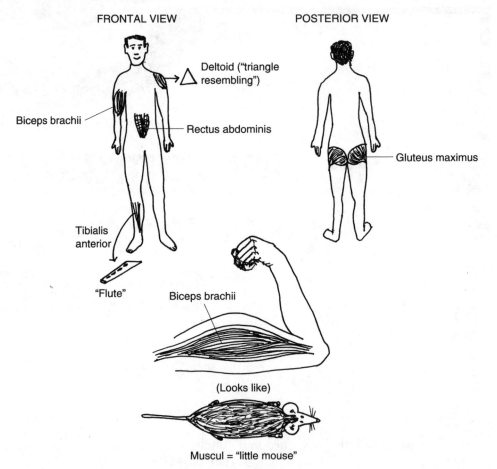

Fig. 14.1 An overview of the human muscular system: A collection of "little mice."

Bone–Muscle Lever Systems

Just as little mice run up-and-down, the skeletal muscles contract and shorten, then relax and lengthen, again. Each of the skeletal muscles is attached to a bone of the skeleton by two or more *tendons* (**TEN**-duns). A tendon is literally a "stretcher." It is actually a strap of dense fibrous connective tissue that anchors a skeletal muscle to a bone. When the muscle contracts, it exerts a pulling force, P. This pulling force stretches the tendon. The tendon then pulls upon the bone.

The human body has a number of *bone–muscle lever* (**LEE**-ver) *systems*. A lever is a rigid bar that is acted upon by some force. In a bone–muscle lever system (Figure 14.2), the bone serves as a passive lever or rigid bar that is pulled upon by a contracting skeletal muscle. The pulling force of the muscle is applied to the bone through the attached tendons.

A *fulcrum* (**FULL**-crumb), abbreviated as F, is a place of "support" upon which a lever turns or balances when it moves. Within a muscle–bone lever system, a movable joint generally serves as the fulcrum. For instance, the biceps brachii contracts and provides a pulling force (P) upon its tendons. The tendons raise the bones of the forearm (the levers) at the elbow joint, which serves as the fulcrum (F). A weight (W) or resistance (R) is then lifted, as the forearm is bent or flexed.

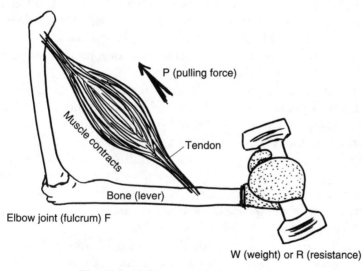

P (pulling force)

Muscle contracts

Tendon

Bone (lever)

Elbow joint (fulcrum) F

W (weight) or R (resistance)

Fig. 14.2   A bone–muscle lever system.

Some Characteristics Used for Naming Muscles

Back in Figure 14.1, you were given anterior (front) and posterior (rear) views of several major muscles of the human body, along with their names. Since this book is entitled *Biology Demystified*, we want to help you take some of the "mystery" out of these muscle names! To do this, we will examine some of the general characteristics or criteria frequently employed to name major skeletal muscles.

Characteristic #1: Body location. Many muscles are either totally or partially named for their body location. Consider, for example, the *tibialis* (**tih-bee-AL**-is) *anterior* muscle. From Figure 14.1, you can see that this muscle is named for its location "in front of" (anterior) to the *tibia* (**TIB**-ee-ah) or "shin bone." The name tibia also means "pipe" or "flute," apparently from its shape. The tibia (and its associated skeletal muscles) is found not just within humans but also in amphibians, reptiles, birds, and other mammals.

Characteristic #2: Muscle shape. Just as the tibia bone (and hence the tibialis anterior muscle) is partially named for its resemblance to an ancient Roman flute, various skeletal muscles are likewise named for their shapes. Look at the *deltoid* (**DELL**-toyd) muscle (Figure 14.1). This muscle "resembles" (*-oid*) a "triangle" (*delt*) in the upper shoulder area.

Characteristic #3: Points of attachment. Remember that each skeletal muscle is attached to a bone by one or more tendons at either end. The *origin* of a muscle is the least movable tendon or attachment of the muscle. Picture it like a heavy anchor. The origin of the biceps brachii muscle, for instance, is at the bones of the shoulder. The *insertion* of a muscle, on the other hand, is the more movable tendon or attachment. The insertion of the biceps brachii is made upon the bones in the lower arm.

2, Order

Within a particular bone–muscle lever system, then, when the muscle contracts and exerts a pulling force, only the insertion end moves. The origin end remains stationary, because it is firmly anchored. During contraction, the insertion end of a muscle moves towards the origin end. In the case of the biceps brachii, when it contracts its lower tendons *flex* (**FLEKS**) or "bend" the arm at the elbow joint. [**Study suggestion:** Put down this book for a moment and flex your own forearm. Try to imagine the lower tendons of your biceps brachii pulling upon their insertion on your forearm bones to raise your lower arm.]

Some muscles are even named for their points of attachment (origin and insertion). A good example is provided by the *sternocleidomastoid* (**ster**-noh-**kleye**-doh-**MASS**-toyd) muscle. Figure 14.3 reveals that this muscle attaches to the *sternum* (**STER**-num) or main bone of the "chest" (*stern*). Its tendons also hook onto the *clavicle* (**KLAV**-uh-kul) or collar bone, which is named for its resemblance to a "little key" (*clavicul*) used in ancient times. Finally, the sternocleidomastoid tendons attach to a third place – the *mastoid* (**MASS**-toyd) *process* of the skull – which literally "resembles" (*-oid*) a "little breast" (*mast*).

When the sternocleidomastoid contracts, it nods the head, drawing the chin down upon the chest. [**Study suggestion:** From the preceding information, which attachment do you think represents the insertion end of the sternocleidomastoid? – What attachments are the origin end?]

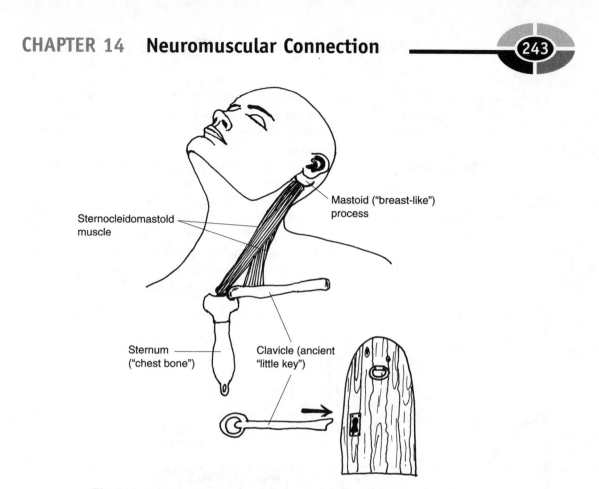

Sternocleidomastoid muscle

Mastoid ("breast-like") process

Sternum ("chest bone")

Clavicle (ancient "little key")

Fig. 14.3   The sternocleidomastoid muscle and its attachments.

Characteristic #4: Relative size. Some muscles are named for their relative size: that is, how big they are compared to their neighbors. Consider two muscles in the *gluteal* (**GLOO**-tee-al) or "rump" region – the *gluteus* (**GLOO**-tee-us) *maximus* (**MACKS**-ih-mus) and the *gluteus minimus* (**MIN**-ih-mus). [**Study suggestion:** Which do you think is the larger muscle, the one with a more "minimum" or more "maximum" size?]

Characteristic #5: Number of muscle "heads." A *muscle head* is a major division of a muscle, with one or more tendons attached. In Latin, *ceps* (**SEPS**) means "head," while *bi-* denotes "two." *Biceps* (**BUY**-seps), therefore, means "two-headed."

Brachii (**BRAY**-kee-eye) stands for "arm." Hence, the name of the biceps brachii muscle translates exactly to mean a "two-headed" muscle in the upper "arm." The biceps brachii has two major heads or divisions to it, and is hooked onto bones by two tendons above the muscle, and two below it.

Characteristic #6: Direction of muscle fibers. Several muscles are given Latin names describing the direction of their individual *muscle fibers* –

long, slender, fiber-shaped muscle cells. Consider the *rectus* (**WRECK**-tus) *abdominis* (ahb-**DAHM**-ih-nus) muscle. (Review Figure 14.1.) It is located in the front of the abdomen (trunk midsection), and its fibers are oriented in a "straight" (*rect*), vertical direction.

Characteristic #7: Association with real or mythological characters. Certain muscles are associated with real people, or with people who only existed in myth. A very intriguing example is the *sartorius* (sar-**TOR**-ee-us). This muscle oddly means "presence of" (*-us*) a "tailor" (*sartori*)! The naming connection is to the way tailors used to sit in ancient times – cross-legged upon the ground (Figure 14.4). The sartorius is actually located along the inner aspect of each thigh. Thus, when it contracts, it flexes (bends) the lower leg, as was done by ancient tailors who used to sit on their bent legs!

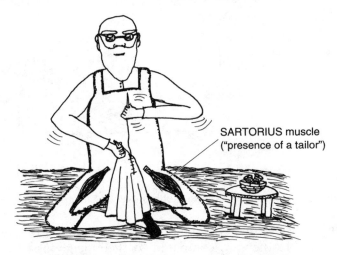

SARTORIUS muscle
("presence of a tailor")

Fig. 14.4 The sartorius: A tale about tailors!

Characteristic #8: Major body actions. Some muscles are named for their major body actions. Consider as an example the *masseter* (mah-**SEE**-ter) muscle in the lower jaw. Its name comes from Latin for "chewer." You use the masseter to lift your lower jaw as you chew meat or gum, of course!

The General Anatomy of a Skeletal Muscle

We have been considering the specific names and appearances of various skeletal muscles. But there is also a general anatomy that practically all skeletal muscles share.

Figure 14.5 includes examples of *fascia* (**FASH**-ee-uh). A fascia is a thin "band" (*fasci*) of fibrous connective tissue surrounding or penetrating a muscle. Consider, for instance, the *epimysium* (**ep**-uh-**MIS**-ee-um), which is a sheet or band of fascia that is "present" (*-um*) "upon" (*epi-*) the entire "muscle" (*mys*). [**Study suggestion:** Look at a raw piece of chicken. Peel the skin back slightly from the flesh, and you will see a milky looking membrane lying upon the meat. What is the name of this membrane?]

Cutting the muscle in half, we can view the *perimysium* (**pair**-uh-**MIZH**-ee-um). The perimysium penetrates deeply into the muscle organ and subdivides it into *fascicles* (**FAS**-uh-kuls) – "little bundles" of muscle fibers that are surrounded by sheets of fascia. The perimysium, therefore, is the fascia present "around" (*peri-*) each bundle or fascicle of muscle fibers.

Finally, the *endomysium* (**en**-doh-**MIZH**-ee-um) is the fascia present "within" (*endo-*) each bundle or fascicle, and between its individual muscle fibers. You will note from Figure 14.5 that the skeletal muscle fibers, themselves, are *striated* (**STRY**-ay-tid) or "furrowed," that is cross-striped with blackish lines.

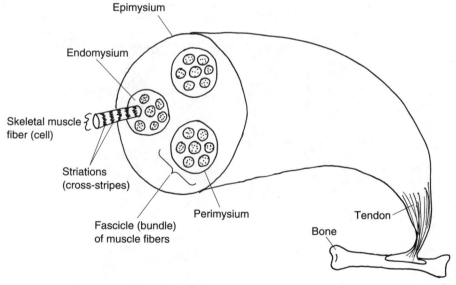

Fig. 14.5   The internal anatomy of a skeletal muscle.

STRUCTURES WITHIN THE MUSCLE FIBERS

An important question for us to ask is, "Okay, but so far we haven't learned how the internal anatomy of a skeletal muscle explains how it contracts (shortens) and provides the pulling force for body movements."

Ultimately, the answer to this question will require us to examine the inner anatomy of an individual skeletal muscle fiber, which is actually a long, fiber-shaped cell.

As Figure 14.6 reveals, the striated (cross-striped) muscle cell or fiber contains numerous *myofibrils* (**my-uh-FEYE**-brils). The word myofibril literally means "little fiber" (*fibril*) of a "muscle" (*my*). Each of these myofibrils is actually a slender, fiber-shaped, cell organelle. The myofibrils have a dark-and-light *banding pattern*. The dark bands are called the *A bands*, while the light bands are called the *I bands*. The *striations* (**stry-AY**-shuns) or cross-stripes of each muscle fiber, then, in reality just represent the dark A bands of their myofibrils, stacked one upon the other to make a stripe.

Within the middle of each light I band is a dark, zig-zagging *Z-line*. These dark lines mark off a series of *sarcomeres* (**SAR**-koh-**meers**). A sarcomere is a short "segment" (*-mere*) of "flesh" (*sarc*): that is, a region of myofibril between two Z-lines. Hence, each myofibril organelle within a muscle fiber basically consists of a series of sarcomeres, attached end-to-end.

3, Order

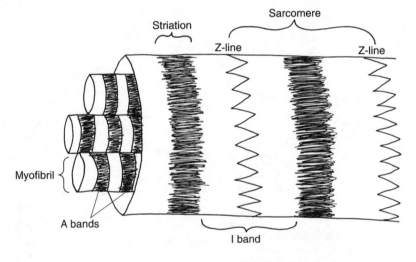

A MUSCLE FIBER (CELL)

Fig. 14.6   A look within a muscle fiber (cell).

Myofilaments and the Sliding Filament Theory

Whether one is considering the contraction of muscles in humans, fish, amphibians, reptiles, birds, or practically any other vertebrate animals, the general mechanism of shortening is much the same. To understand its basic features, we must examine the interior of a sarcomere (Figure 14.7). Within the sarcomere are a number of *myofilaments* (**my**-oh-**FIL**-ah-ments) or "muscle threads." In reality, these myofilaments are thread-like collections of protein molecules. Attached to the Z-line at either end of the sarcomere are a series of *thin actin* (**AK**-tin) *myofilaments*. Each of these thin actin myofilaments consists of two twisted strands of globe-shaped actin proteins. [**Study suggestion:** Find a beaded pearl necklace and lay it upon a surface. Place both strands of the necklace side-by-side, then twist them around each other. The resulting double helix (two twisted strands) provides a rough model for a thin actin myofilament.]

In the middle of each sarcomere is a series of *thick myosin* (**MY**-oh-sin) *myofilaments*. These are stacked vertically above-and-below one another, with narrow gaps between them. Each myosin myofilament consists of dozens of individual myosin protein molecules. Each myosin protein somewhat resembles a golf club with two heads. The double-heads are tiltable, as if they were poised upon a chemical hinge. The tiltable double-head of each myosin molecule is technically called a *myosin cross-bridge*. [**Study suggestion:** Visualize two golfers, each carrying their own bag of golf clubs. Each club has a double-head at one end, which is attached by a tiltable hinge. The two golfers stand back-to-back, in the center of a sarcomere, and then each gives his golf bag a heave. If the golfers keep hold of their bags, their clubs will come flying out in both directions, some with their double-heads pointing upward, and some with their double-heads pointing down. The resulting highly orderly arrangement provides a rough model for the thick myosin myofilament.]

The myosin cross-bridge is the chief contact point between the thin actin and thick myosin myofilaments. It is also vitally important because of its close functional relationship with the high-energy ATP molecule. You may remember (Chapter 4) that the ATP molecule is split by a special kind of enzyme, called ATPase. In the case of muscle, the enzyme is *myosin ATPase*.

THE SLIDING FILAMENT THEORY

According to the *sliding filament theory of muscle contraction*, muscles shorten due to the inward sliding of the thin actin myofilaments over the tilted cross-bridges of the thick myosin myofilaments. "How is this accomplished?" you might reasonably question. The answer involves the action of myosin ATPase.

4, Order

When the skeletal muscle fiber is excited by a nerve ending, myosin ATPase enzyme splits ATP molecules in the region of the myosin cross-bridges. The resulting free energy tilts the myosin cross-bridges inward. As the cross-bridges tilt, the overhanging thin actin myofilaments slide over their tips. The sliding occurs at both ends of the sarcomere. Hence, each sarcomere shortens. Since each myofibril organelle consists of a series of sarcomeres hooked end-to-end, the whole myofibril shortens. And as all of their myofibril organelles shorten, the entire skeletal muscle fiber (cell) also shortens.

It is this shortening of many muscle fibers that yanks upon a tendon, creating a pulling force upon a bone that results in body movement.

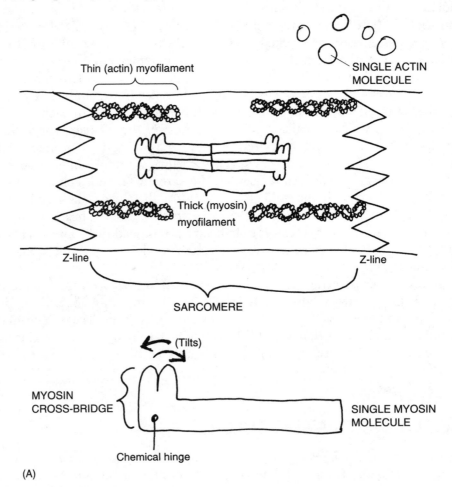

Fig. 14.7 Myofilaments and the sliding filament theory of muscle contraction. (A) Resting (non-contracting position). (B) Contracting process: thin myofilaments slide inward; sarcomere shortens.

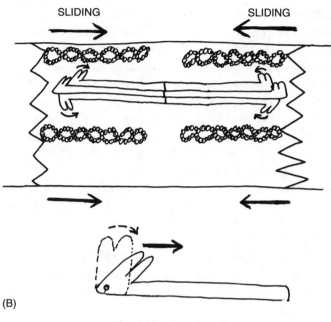

SLIDING   SLIDING

(B)

Fig. 14.7   (continued)

After the muscle contracts, the cross-bridges tilt back into their vertical positions, causing the thin actin myofilaments to slide outward, again. The sarcomere re-lengthens, and the muscle fiber relaxes. This same highly orderly sequence of muscle contraction (shortening) followed by relaxation (lengthening) repeats itself over and over again. And each time, the trigger for the contraction–relaxation sequence to begin is a sufficiently strong level of muscle fiber excitation by nearby nerve endings.

The Neuromuscular Junction and Muscle Excitation

Remember that this chapter is entitled "The Neuromuscular (Nerve–Muscle) Connection." A critical element in this connection is the *motor neuron*. As its name suggests, a motor neuron is a nerve cell that excites a muscle to contract (or a gland to secrete), thereby causing movement. Thus, this type of neuron results in the moving of a vertebrate body, in somewhat the same sense that a motor ultimately causes the body of a car to move.

Figure 14.8 illustrates the main parts of a typical motor neuron. There are a number of *dendrites* (**DEN**-dryts) at one end of the nerve cell. These dendrites are actually slender branches of cytoplasm that carry excitation toward the *cell body* (major central portion) of the neuron. [**Study suggestion:** Why do you think this word derives from the Greek for "of a tree"?]

When the cell body is sufficiently excited, it fires off an *action potential* or *nerve impulse*. The action potential (nerve impulse) can most simply be considered a wave. It is a traveling wave of chemical excitation that passes from the neuron cell body, and then onto the neuron *axon* (**AX**-ahn).

The axon is like a long slender "axle" holding two car wheels. Although the axon doesn't move, it is a long, slender branch of cytoplasm that carries excitation (an action potential) away from the cell body of a neuron. The action potential or nerve impulse travels down the axon to its branching tips, called the *axon terminals*.

Within the axon terminals are tiny *vesicles* (**VES**-ih-kls). Each vesicle is literally a "tiny bladder" that consists of a membrane surrounding thousands of *neurotransmitter* (**NUR**-oh-**trans**-mit-er) *molecules*. Each time an action

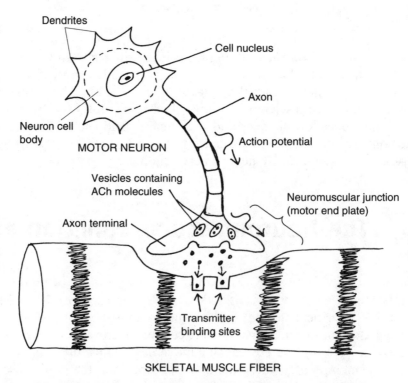

Fig. 14.8   The neuromuscular junction and muscle excitation.

potential travels down into the axon terminals, the membranes surrounding some of the vesicles rupture.

Hundreds of neurotransmitter molecules are then released into the neuromuscular junction, which is also called the *motor end plate*. The neuromuscular junction (motor end plate) is the flat, plate-like area where the axon terminals of a motor neuron almost (but not quite) touch the cell membrane of a muscle fiber.

THE PROCESS OF MUSCLE FIBER EXCITATION

The released neurotransmitter molecules diffuse across the narrow, saltwater-filled gap of the neuromuscular junction, then attach to *transmitter binding sites* on the muscle fiber. What happens to the muscle fiber, then, is largely determined by the nature of the neurotransmitter molecules that are released and attached to the binding sites.

Excitatory neurotransmitter molecules stimulate or excite the skeletal muscle fiber to contract. The main excitatory neurotransmitter in the human neuromuscular junction is called *acetylcholine* (uh-**see**-tul-**KOH**-leen), abbreviated as *ACh*.

Motor Pathways

The neuromuscular junction (motor end plate) can be considered the functional end of a particular *motor pathway*. A motor pathway is a sequence of one or more motor neurons that carries an action potential to a neuromuscular junction, thereby resulting in muscle contraction.

From this definition, you can see that a *lower motor neuron* is the one whose axon and axon terminals actually makes a neuromuscular junction with one or more skeletal muscle fibers. "Where is the cell body of this lower motor neuron located?" the inquiring mind needs to know. Quite often, the cell body of the lower motor neuron is located in the *gray matter* of the *spinal cord* (Figure 14.9). The axon then extends outward into the *white matter*, entering a *spinal nerve*.

The spinal nerve travels some distance away from the spinal cord. The motor neuron axons within the spinal nerve carry the action potential away from the central region of the whole body, and hence to the "outer portion" or *periphery* (per-**IF**-eh-ree). As it reaches the body periphery, the spinal nerve branches into one or more *peripheral* (per-**IF**-er-al) *nerves*. It is the peripheral nerve, then, that forms the last major linkage in the motor pathway. From here, the axon terminals of the lower motor neuron branch out and make a neuromuscular junction with one or more skeletal muscle fibers. Excitatory neurotransmitter (such as acetylcholine, ACh) is released, the skeletal muscle fibers are stimulated, and the fibers contract.

The spinal nerves and individual peripheral nerves leading to various skeletal muscles are considered parts of the *Peripheral Nervous System* or *PNS*. This name is due to the fact that they supply the body periphery (outer portion).

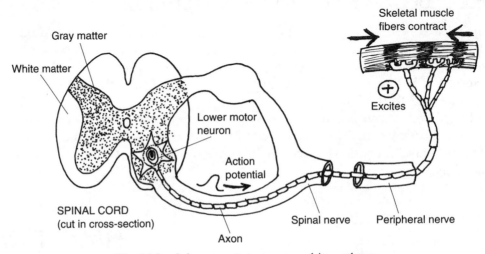

Fig. 14.9   A lower motor neuron and its pathway.

HIGHER MOTOR NEURONS WITHIN THE CNS

We have explained how a lower motor neuron in the gray matter of the spinal cord can excite skeletal muscle fibers to contract. The next question to consider is, "What is it that excites a lower motor neuron, so that it can then excite skeletal muscle?"

The answer is that a series of one or more *higher motor neurons* within a particular motor pathway eventually stimulates a lower motor neuron. The cell bodies of such higher motor neurons generally lie within the *Central Nervous System*, which is abbreviated as *CNS*. The Central Nervous System (CNS) is the portion of the nervous system that is centrally located within the skull and the vertebral column. Specifically,

$$\text{Central Nervous System} = \text{Brain} + \text{Spinal cord}$$
$$\text{(CNS)}$$

The *brain* is technically the upper portion of the CNS that lies within the skull. The spinal cord, in contrast, is the narrow, "cord"-like lower portion of the CNS that is housed within the vertebral column.

Figure 14.10 provides an overview of the CNS, along with the relative positions of the brain and spinal cord. One especially important location for higher motor neurons in the brain is an area called the *precentral gyrus* (**JEYE**-rus). The name gyrus means "ring" or "fold." The precentral gyrus, therefore, is a raised ring or fold of brain tissue located just "before" or "in front of" (pre-) a groove called the *central sulcus* (**SUL**-kus).

The precentral gyrus lies at the superior end of the *cerebrum* (seh-**REE**-brum) or "main brain mass." Since this gyrus is at the surface, it is part of the *cerebral cortex* (**KOR**-tex) or gray matter "bark" covering the "main brain

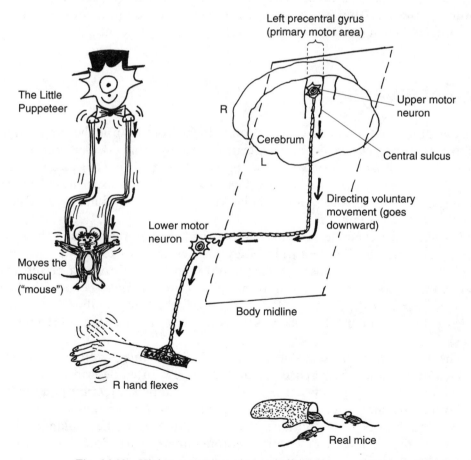

Fig. 14.10 Higher motor neurons and the hidden puppeteer.

mass." The upper motor neurons situated within the cerebral cortex of the precentral gyrus often direct voluntary movements of the body to occur.

The precentral gyrus is sometimes called the *primary motor area*. It is here that upper motor neurons "decide" to initiate the movements of various skeletal muscles (usually on the opposite side of the body). From the primary motor area (precentral gyrus), motor nerve fibers descend, cross over the *body midline*, and then stimulate skeletal muscles on the opposite side of the body to contract. To move your right hand, for instance, upper motor neurons within the left precentral gyrus would send down a message, which would eventually cross to the right side of the spinal cord, then out to the muscles in the right hand.

[**Study suggestion:** Imagine a mischievous little puppeteer hiding within your left precentral gyrus. The strings manipulating his puppet hang down (descend) for some distance. These "strings" are actually nerve fibers in a motor pathway. After the strings are pulled (excited by a muscle action potential), they activate the puppet (skeletal muscle) to contract.]

Sensory Pathways

Just as motor pathways descend or come down from both higher and lower motor neurons, *sensory pathways* ascend or rise up from *sensory neurons*. A *lower or first-order sensory neuron* is a nerve cell that has part of its dendrites modified to form *sensory receptors*. Sensory receptors are specialized nerve endings that are sensitive to a particular kind of "goad" (poking prod) called a *stimulus* (**STIM**-yuh-les).

Consider, for example, certain onion-shaped sensory receptors in the skin of the right hand (Figure 14.11). They are sensitive to touch or pressure sensations. A nerve impulse is created, and then it travels over a peripheral nerve, then a spinal nerve, and into the spinal cord. A *higher or second-order sensory neuron* is then excited. Its axon crosses the body midline to the left side of the spinal cord, and then ascends towards the thalamus (**THAL**-uh-**mus**).

The thalamus is an egg-shaped "bedroom" (*thalam*) – an oval region nestled deep within the cerebrum. The thalamus is often described as a *sensory relay center*. This is because sensory nerve impulses (action potentials) are often switched over (or relayed) up onto *third-order sensory neurons*, whose cell bodies lie within the thalamus. And from the thalamus, the axons of the third-order sensory neurons finally carry the information about the stimulus all the way up to the *postcentral gyrus*.

As its name clearly indicates, the postcentral gyrus is a raised ring or fold of cerebral cortex that runs up and down just behind or "after" (*post-*) the central sulcus. The postcentral gyrus is alternately called the *primary or general sensory cortex*. This name reflects the fact that the postcentral gyrus area is the final destination for most of our general or primary body senses, such as touch, pressure, temperature, vibration, and pain.

Remember that we imagined there was a hidden puppeteer within the left precentral gyrus, directing much of the motor pathway descending on the right side of the body. Likewise, we can draw a *sensory homunculus* (hoh-**MUNG**-kyuh-lus) – a "little feeling man or dwarf" (*homuncul*) – just above the left postcentral gyrus. This little feeling dwarf is a rather bizarre visual method by which anatomists have commonly mapped the final destinations of sensory nerve fibers ascending from various parts on the right side of the body.

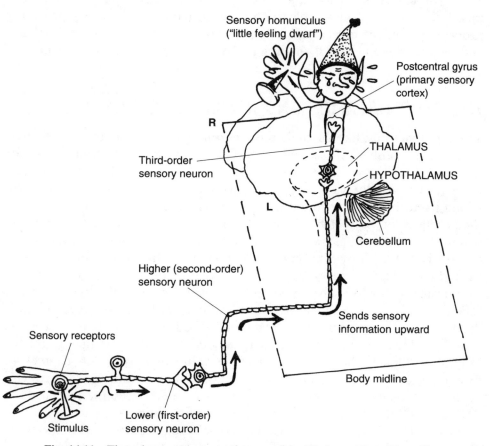

Fig. 14.11   The primary sensory pathway and its "little man" or "dwarf".

The Cerebellum and Hypothalamus

The thalamus and postcentral gyrus (primary or general sensory cortex) are not the only places in the human brain that receive sensory information. Two other very important sensory areas are the *cerebellum* (**sair**-uh-**BELL**-um) or "little cerebrum" (*cerebell*) and the *hypothalamus* (**high**-poh-**THAL**-uh-mus) or region "below" (*hypo*-) the "thalamus." (Review Figure 14.11.)

The cerebellum is involved in *proprioception* (**proh**-pree-oh-**SEP**-shun). This long term literally means "the process of receiving" (-*ception*) sensations from "one's own" (*propri*) self. In more direct language, proprioception involves the receiving of sensory stimuli that indicate the relative position of the entire body and its parts within space. The sensory receptors for proprioception are often found within the skeletal muscles and joints. Thus, you can sense or feel that you are standing upright, or that you have raised your arm up off a table, without even having to look! By providing proprioception, the cerebellum helps the body maintain its vertical balance and reduces the amount of shaking during automatic reflex movements.

THE HYPOTHALAMUS AND NEGATIVE FEEDBACK

The hypothalamus, like the thalamus, lies buried deep within the cerebrum. It contains a number of *control centers* for homeostasis. It includes, for instance, the *temperature control center*. Recall (Chapter 1) that oral body temperature usually varies up-and-down in a roughly S-shaped pattern, within its normal range. And remember (Chapter 13) how thermoregulation (homeostasis of body temperature) is maintained by processes occurring within the skin, such as sweating and vasodilation. The logical question we can ask is, "Okay, just what part of the body is it, that controls the mechanisms responsible for carrying out thermoregulation?" The answer is this: Certain neurons within the temperature control center of the hypothalamus establish and maintain the *set-point* for oral body temperature. A set-point is defined as the long-term average value of a body variable: that is, it is the *point* at which the variable seems to be *set*. Oral body temperature, for example, seems to be set at an average value of about 98.6 degrees F. [**Study suggestion:** Visualize the thermostat in your own house or apartment. What is the usual "set-point" temperature where you keep the thermostat?]

Obviously, when certain factors raise the body temperature too far above this set-point (or too far below it), then some type of *control system* has to be engaged. Usually, the human body uses a *negative feedback control system* to correct or minimize the amount of change.

5, Order

In general, a negative feedback control system is a collection of anatomical and physiological components that operate to produce an outcome or response which curves or feeds back upon the initial stimulus in a negative manner. By this it is meant that the final outcome of the system curves (feeds back) upon the initial stimulus to remove or correct it. (Consult Figure 14.12, A.)

Consider, for instance, a situation where the temperature in a room starts to rise above its set-point value on the thermostat (Figure 14.12, B). As the temperature begins to rise towards the upper limit of its normal range, a negative feedback temperature control system kicks in. The rise in temperature acts as a stimulus for the air conditioner in the unit. It begins to blow cold air into the room. Soon, the room temperature drops back down towards its set-point value on the thermostat. The intial stimulus (rise in

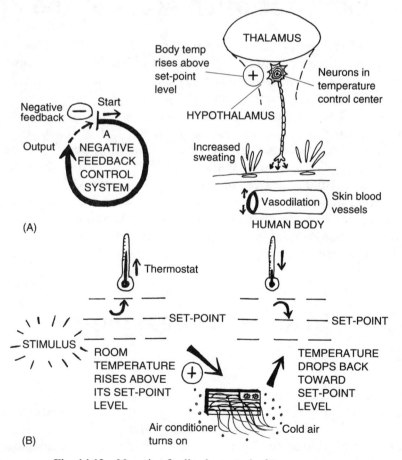

Fig. 14.12   Negative feedback control of temperature.

room temperature above the set-point level) has therefore been corrected or removed. Hence, a negative feedback control system has been in operation.

Within the human body, of course, increased sweating and vasodilation of skin blood vessels would have a similar negative feedback effect upon any significant rise in oral body temperature beyond its set-point of about 98.6 degrees F. A group of neurons within the temperature control center of the hypothalamus direct this sweating and vasodilation process to occur.

In addition to the temperature control center, the hypothalamus has numerous other control centers. These include a *hunger center*, *thirst center*, *pleasure center*, and many other centers that help control some important body variable and thus maintain homeostasis.

Integration and Association Areas of the Brain

In addition to the major motor and sensory areas of the Central Nervous System (brain and spinal cord), there are a number of critical *sensory integration or association areas*. These areas make meaningful connections or associations between incoming stimuli, so that recognition and learning tend to occur. Take, for instance, the *primary visual area* at the back of the cerebrum, and the *visual association area* located just in front of it (Figure 14.13). The primary visual area receives visual impulses from the *retina* (**RET**-ih-nah) in the rear wall of the eyeball. These visual impulses travel to the primary visual area over the *optic* (**AHP**-tik) *nerves*. But these visual impulses don't make any sense to us until they are sent forward to the visual association area. Here the separate bits of visual information are ordered and assembled into meaningful patterns and images that we can recognize, such as the familiar face of a good friend.

Neuromuscular Disorders: Breaking the Connections

We have now outlined some of the major aspects of the nervous and muscular systems. Primarily, this outline has involved basic features of their normal anatomy and physiology. To a large degree, these two systems are functionally linked due to the extremely high degree of Biological Order found in the connections between them. Hence, neuromuscular disorders,

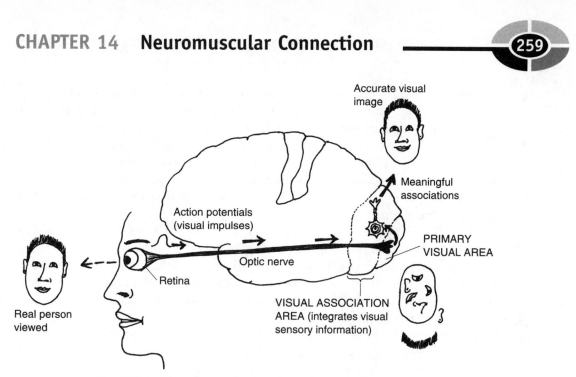

Fig. 14.13 The primary visual and visual association areas.

and the disease/injury states often associated with them, frequently result from the breaking or disruption of the normal connections between muscles and nerves.

1, Disorder

PARALYSIS AND DENERVATION ATROPHY

Consider what happens, for instance, when part of the major motor pathway descending to the skeletal muscles is *transected* (tran-**SEK**-ted) or "cut" (*sect*) all the way "through" (*trans-*). When the motor nerve fibers from the precentral gyrus (primary motor area) are transected (cut all the way through), say, from a terrible car accident, then either partial or full *paralysis* (puh-**RAL**-uh-sis) of the body occurs. Paralysis is literally a state of "disablement," an inability to move. The axons of the motor neurons leading from the motor areas of the cerebral cortex are unable to complete their final connection with the neuromuscular junction (motor end plate).

2, Disorder

As a result, there is either a full or total paralysis of the skeletal muscles being supplied. [**Study suggestion:** If the main motor pathway leading down from the left precentral gyrus is transected, then many of the skeletal muscles on which side of the body will be paralyzed? Why?]

Troph (**TROHF**) is a root that means "nourishment." But troph means nourishment in its broadest sense, including "stimulation." The skeletal muscle fibers, then, receive "nourishment" by the motor nerve fibers coming down to the neuromuscular junction, because they are stimulated to contract by them. In technical terms, we say that the skeletal muscle fibers are *innervated* (**IN**-ur-**vayt**-ed) or have "nerve" (*nerv*) endings put "into" (*in*-) them. This situation of *innervation* (in-er-**VAY**-shun) essentially occurs at the neuromuscular junction. The skeletal muscle fibers are therefore innervated at the neuromuscular junction and receive nourishment or stimulation (troph) from them.

Denervation (**dee**-ner-**VAY**-shun) is the "process of" (*-tion*) taking "nerves" (*nerv*) "away from" (*de*-) muscle fibers. If some part of the motor nerve pathway undergoes transection, then the skeletal muscle fibers ultimately being served by the pathway will suffer denervation. They will also undergo a type of *atrophy* (**AT**-ruh-**fee**) – a "condition" (*-y*) "without" (*a*-) "nourishment" (troph) or stimulation.

The complete phrase is denervation atrophy, which means the wasting away and degeneration of skeletal muscle fibers that have been deprived of their normal stimulation or nourishment by nerve endings. The *denervated* (**DEH**-ner-**vayt**-ed) muscles become *flaccid* (**FLAH**-sid) or "flabby" and almost useless.

ANESTHESIA – LOSS OF FEELING OR SENSATION

3, Disorder

When the sensory nerve fibers leading from sensory receptors in muscles, joints, skin, or elsewhere are transected, then *anesthesia* (**an**-es-**THEE**-zhuh) often results. Anesthesia is a "condition" (*-ia*) "without" (*an*-) "feeling or sensation" (*esthes*).

If, for instance, the sensory nerves leading from most of the receptors in your big toe are completely transected, then the big toe is pretty much numb, reflecting a condition of anesthesia. [**Study suggestion:** Assume that the main sensory nerve pathway ascending from your right big toe is completely transected. Going back and reviewing past figures in this chapter, which specific gyrus area receiving big toe sensations would be deprived? On which side of the body midline would it most likely be located?]

Quiz

Refer to the text in this chapter if necessary. A good score is at least 8 correct answers out of these 10 questions. The answers are listed in the back of this book.

1. The muscular system of humans:
   (a) Involves mainly the muscles in the walls of the heart
   (b) Is chiefly composed of over 600 individual skeletal muscle organs
   (c) Focuses upon about 700 pairs of skeletal muscle tissues
   (d) Bears no relationship to bones or tendons

2. The masseter muscle in the jaw is named for the characteristic of:
   (a) Muscle shape
   (b) Body location
   (c) Major body action
   (d) Number of muscle heads

3. In bending and straightening of the lower leg, the knee joint serves as a:
   (a) Fulcrum
   (b) Lever
   (c) Resistance
   (d) Pulling force

4. When the biceps brachii contracts, it flexes (bends) the lower arm at the elbow. The insertion of the biceps brachii must therefore be at the:
   (a) Shoulder
   (b) Wrist
   (c) Fingers
   (d) Elbow

5. Epimysium represents:
   (a) A special type of joint
   (b) A type of fascia located upon an entire skeletal muscle
   (c) Fascia located around a bundle of muscle fibers
   (d) Striated muscle fibers

6. The striations of an individual skeletal muscle fiber are chiefly due to:
   (a) A stack of light I bands within the myofibrils
   (b) Several groups of muscle cell nuclei closely adjacent to one another
   (c) A stack of dark A bands within the myofibrils
   (d) Complete lack of any significant organelle banding patterns

7. The thin myofilaments primarily consist of the protein ____
   (a) Actin
   (b) Myosin
   (c) Z-line
   (d) Cross-bridge substance

8. According to modern theory, a muscle fiber contracts because:
   (a) Its thin myofilaments become shorter
   (b) The relaxation of the muscle is no longer being opposed
   (c) The thick myofilaments lengthen considerably
   (d) The thin myofilaments slide inward over the tilted cross-bridges of myosin

9. The neuromuscular junction is alternately called the:
   (a) Hypothalamus
   (b) Motor end plate
   (c) Axon terminals
   (d) Vesicles

10. Many sensory nerve fibers ascend and relay to the cerebral cortex, and its:
    (a) Postcentral gyrus
    (b) Thalamus
    (c) Hypothalamus
    (d) Primary motor area

The Giraffe ORDER TABLE for Chapter 14
(Key Text Facts About Biological Order Within An Organism)

1. _____
2. _____
3. _____
4. _____
5. _____

The Dead Giraffe DISORDER TABLE for Chapter 14
(Key Text Facts About Biological Disorder Within An Organism)

1. _____

2. _____

3. _____

Glands and Their Hormone Messengers

Chapter 14 outlined the neuromuscular system. It included a description of the nerve impulse or action potential, a traveling wave of chemical excitation. Remember that this traveling wave allows a motor neuron to stimulate skeletal muscle fibers at the neuromuscular junction, and it permits one neuron to communicate with other neurons across a *synapse* (**SIN**-aps). A synapse is literally the place where two neurons come very close (without touching), so that they almost "fasten together."

Two Solutions to Communication: Neurotransmitters versus Secretions

1, Order

The synapse between two neurons, like the neuromuscular junction between a motor neuron and a muscle fiber, employs a special type of chemical communication (see Figure 15.1, A and B). Specifically, vesicles within the axon terminals of an excited neuron rupture and release neurotransmitter molecules (such as acetylcholine). The neurotransmitter molecules then diffuse and bind to the membrane of either another neuron (in the case of a synapse) or a skeletal muscle fiber (in the case of a neuromuscular junction). In both cases, however, there is a type of chemical communication between cells that occurs with the release of neurotransmitter molecules.

GLANDS AND SECRETION

But there is another mode or way of chemical communication between cells. This mode is called *secretion* (sih-**KREE**-shun). Secretion is literally the "process of separating" – the separation of certain substances from the bloodstream, followed by their release.

The *secreted* (see-**KREE**-ted) substance usually comes from epithelial cells. Remember (Chapter 2) that epithelial tissue is the body's main covering and lining tissue. But it also occurs within *glands*. A gland is a "little acorn" – a rounded, somewhat "acorn"-shaped mass of one or more epithelial cells that have become specialized for the function of secretion.

2, Order

Exocrine (**EK**-suh-krin) *glands* are glands of "external" (*exo-*) secretion that release some useful product into a duct, which then carries the secretion to some body surface (study Figure 15.1, C). A good example of exocrine glands are the sweat glands of the skin, which secrete sweat into numerous sweat ducts. Sweat, as you no doubt remember, is a useful product because it helps cool the body and prevent hyperthermia (excessive body temperature).

Endocrine (**EN**-doh-krin) *glands*, in contrast, are glands of "internal" (*endo-*) secretion that release a *hormone* into the bloodstream, right within the gland itself. A hormone is literally "an arouser" (*hormon*). A hormone is a chemical messenger secreted into the bloodstream by an endocrine gland. It gets its name from the fact that the hormone often "arouses" (stimulates) certain *target cells* in the body to increase their activity. The target cells of the hormone may be located far downstream, but the blood will eventually circulate to bring the hormone molecules to them (see Figure 15.1, D).

3, Order

The Neuroendocrine System

The endocrine glands, like the nervous system, are vitally important in both communication and control of the internal environment within the bodies of

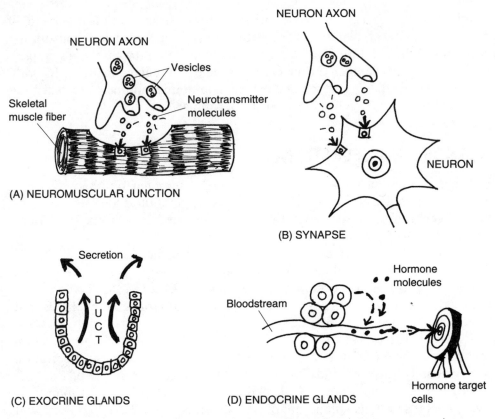

Fig. 15.1 Two ways for cells to communicate: Neurotransmitters versus secretions.

4, Order

vertebrates. In fact, it is sometimes very hard to separate the nervous and endocrine systems at all! In these cases, we use the term *neuroendocrine* (**NUR**-oh-**en**-doh-krin) *system* to describe them. The neuroendocrine system is an organ system that contains parts of the nervous system, as well as parts of the endocrine gland system. Thus, communication can occur via release of neurotransmitters (the nervous component), and also via secretion of particular hormones into the bloodstream (the endocrine component). We will now describe some specific examples.

THE HYPOTHALAMUS–PITUITARY CONNECTION

Chapter 14 described the hypothalamus as an area deep within the cerebrum that contains a number of control centers (such as the temperature control center) that are vitally important for maintaining various aspects of home-

ostasis. Located just below the hypothalamus is a narrow, funnel-shaped
pituitary (pih-**TWO**-eh-**tear**-ee) *stalk*. This hollow stalk is attached to the
top of the *pituitary body*. The front of the pituitary body is called the *anterior
pituitary* gland, while the back is called the *posterior pituitary* gland. The
exact Latin translation of the word pituitary becomes evident from a look
at Figure 15.2. It literally "pertains to *mucus* (**MYOO**-kus) or *phlegm* (flem)."
This term derives from the fact that the pituitary body is located just above
and behind the nose. Hence, it was only logical for the ancient anatomists to
think that the pituitary was the source of the "slime" (mucus) that sometimes
leaks from your nose when its interior is "inflamed" (phlegm)!

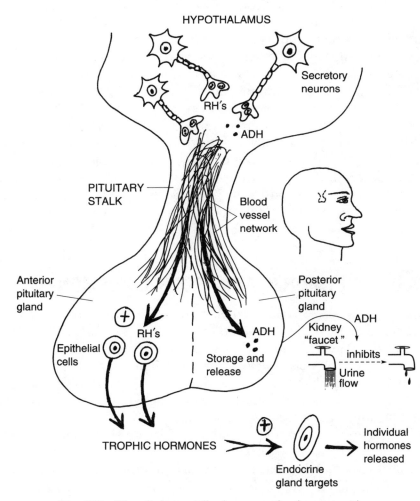

Fig. 15.2   The pituitary stalk: A neuroendocrine connection.

Even though we dwellers of the 21st century realize that the pituitary body is not some kind of channel that carries mucus from the brain down into the nose, we do recognize that there is definitely a *hypothalamus–pituitary connection*. The main connection is through the pituitary stalk. But instead of carrying mucus or phlegm, the pituitary stalk contains a network of tiny blood vessels that run down from the hypothalamus into the pituitary body.

In addition to vital control centers, the hypothalamus also contains an amazing collection of *secretory* (**SEE**-kreh-**tor**-ee) *neurons* – neurons that secrete hormones or hormone-like substances. These hormones (or hormone-like substances) are secreted into the tiny blood vessels flowing down through the pituitary stalk. Thus, the pituitary stalk (and its collection of blood vessels) creates an important neuroendocrine connection. This is because the hypothalamus is generally considered part of the nervous system, while the pituitary body and its two glands (anterior pituitary and posterior pituitary) are classified as members of the large endocrine gland system.

One group of secretory neurons produce *antidiuretic* (**an**-tee-**die**-yuh-**RET**-ik) *hormone*. Antidiuretic hormone, abbreviated as *ADH*, is named for its primary function. ADH helps the kidney retain water due to its effect "against" (*anti-*) too much "urine passing through" (*diuret*) and out of the body. Although ADH is actually secreted by the neurons in the hypothalamus, the hormone molecules travel down through the blood vessels in the pituitary stalk, and are then stored by the posterior pituitary gland and later released into the general bloodstream. [**Study suggestion:** Look at the "kidney faucet" model in Figure 15.2. How does the definition of ADH help explain the turning off of the faucet from full blast to a slow drip?]

RELEASING HORMONES AND TROPHIC HORMONES

The largest group of secretions coming from the hypothalamus have no direct connection to ADH. These other secretions are called the *releasing hormones (RHs)*. The releasing hormones (RHs) are secreted into the network of tiny brain blood vessels coursing down through the pituitary stalk. The transported RHs finally enter the anterior pituitary gland (rather than the posterior pituitary gland). The releasing hormones get their name from their primary function. They literally stimulate the *release* of a variety of *trophic* (**TROHF**-ik) *hormones* from the epithelial cells of the anterior pituitary gland.

Remember (Chapter 14) that troph means "nourishment" in the sense of stimulating something. Trophic hormones, therefore, are hormones secreted by the anterior pituitary gland that have individual endocrine glands as their main targets for "nourishment" or stimulation. Figure 15.2 outlined the highly orderly three-step process:

1. A particular releasing hormone (RH) is put out by the hypothalamus.
2. Each RH passes down through the pituitary stalk and triggers the release of a certain trophic hormone.
3. Each trophic hormone in turn enters the general bloodstream and stimulates a particular individual endocrine gland target to secrete its own hormone.

Figure 15.3 shows pictures of the major endocrine gland targets being stimulated by various trophic hormones from the anterior pituitary. Take, for instance, the *thyroid* (**THIGH**-royd) *gland* in the front of the neck. The thyroid gland is named for its "resemblance" (-*oid*) to a broad "shield" (*thyr*). The thyroid is stimulated or "nourished" by *thyroid-stimulating hormone*, abbreviated as *TSH*. The thyroid gland is stimulated by TSH to

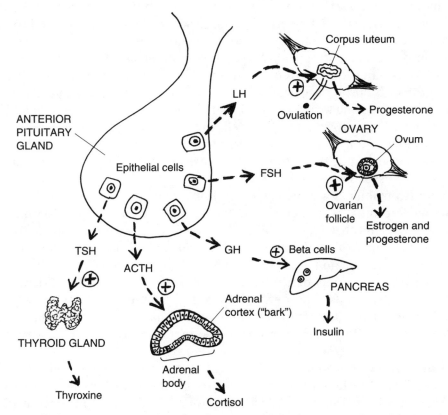

Fig. 15.3 Some major trophic hormones and their endocrine targets.

increase the rate of secretion of its own individual hormone, *thyroxine* (**thigh-ROCKS**-in). Thyroxine, in turn, circulates throughout the bloodstream to affect most of the body cells. Thyroxine, for example, increases the *basal* (**BAY**-sal) *metabolic rate* or *BMR*: that is, the rate at which body cells burn calories during their metabolism under resting or "basal" conditions. [**Study suggestion:** From its effect upon BMR, how would you expect thyroxine to influence oral body temperature? Why?]

The name of a second important trophic hormone is a real tongue-twister! Its name is *adrenocorticotrophic* (uh-**dree**-noh-**kor**-tuh-koh-**TROHF**-ik) *hormone*, simply abbreviated as *ACTH*. The *adrenocortico-* part of the hormone's name comes from its endocrine gland target – the *adrenal* (uh-**DREE**-nal) *cortex*. Just as the cerebral cortex (Chapter 14) literally forms a thin "bark" over the surface of the cerebrum, the adrenal cortex is an endocrine gland forming a thin "bark" (*cortex* or *cortico-*) over the surface of the *adrenal body*. This body is a curved, stocking cap-shaped structure that lies "toward" (*ad-*) the top of each "kidney" (*renal*).

The adrenal cortex or thin outer bark of the adrenal body secretes the hormone *cortisol* (**KOR**-tih-**sol**). Cortisol raises the blood glucose level whenever it is low, and it also acts to relieve the symptoms of tissue inflammation.

A third trophic hormone is called *growth hormone (GH)*. Growth hormone, as its name states, circulates to most of the body cells and stimulates their growth by promoting such processes as protein synthesis and cell division. But the specific target gland of its trophic influence are the *beta* (**BAY**-tuh) cells of the *pancreas*. The beta cells are well known because they secrete the hormone, *insulin* (**IN**-suh-lin). Insulin is absolutely critical for human survival, because it helps to transport glucose out of the bloodstream, thereby feeding the tissue cells.

A fourth trophic hormone is *follicle-stimulating hormone (FSH)*. You may bring back to mind (Chapter 13) the "little bags" or hair follicles within the skin. There is another type of "little bag" or follicle within the *ovaries* (**OH**-var-**ees**) or "eggs" (*ova*) of females. The *ovarian* (oh-**VAIR**-ee-an) *follicles* are tiny bags or sacs within the ovaries. Under the stimulating effect of FSH, they increase their secretion of the two hormones, *estrogen* (**ES**-troh-jen) and *progesterone* (proh-**JES**-ter-**own**). Estrogen stimulates the development of so-called *secondary sex characteristics* in the female, such as a higher voice and softer skin. Progesterone prepares the female body for a possible pregnancy.

A fifth trophic hormone, *luteinizing* (**LEW**-tuh-**neye**-zing) *hormone* or *LH*, also acts upon the female ovary. It triggers a rupture of the mature ovarian follicle, thereby causing *ovulation* (**ahv**-you-**LAY**-shun) – the release of a "little egg" (*ovul*) into the abdominal cavity. But luteinizing hormone derives its name from the *corpus* (**KOR**-pus) *luteum* (**LEW**-tee-um) or "yellow body" that is still left within the ovary after ovulation. LH thus indirectly has a luteinizing or "yellowing" effect, creating a yellowish body (corpus luteum) that secretes lots of progesterone (and some estrogen).

THE SYMPATHETIC NERVE–ADRENAL CONNECTION

In addition to the hypothalamus–pituitary connection, there are other examples of neuroendocrine relationships. Prominent among these are the partners in the "Fight-or-Flight" Stress Response. Whenever a human or other vertebrate is under severe stress, then this "Fight-or-Flight" Response automatically engages. The vertebrate gets ready to either stand its ground and fight for survival, or else take flight and simply run away from encroaching danger.

5, Order

The *sympathetic nerves* are critical elements of the "Fight-or-Flight" Response, as are both the adrenal cortex and the *adrenal medulla* (meh-**DEW**-lah). The adrenal medulla lies in the "middle" (*medull*) of the adrenal body, whereas the adrenal cortex forms a thin bark around it. Figure 15.4 illustrates how the sympathetic nerves leave the hypothalamus area and carry nerve impulses (action potentials) down into the adrenal medulla. The sympathetic nerves are literally the "suffering" (*path*) "with" (*sym-*) nerves, reflecting their central role in carrying out the body's response to alarm or stress.

The sympathetic nerve endings stimulate epithelial cells within the adrenal medulla to increase their secretion of the two related hormones, *epinephrine* (**ep**-ih-**NEF**-rin) and *norepinephrine* (**NOR**-ep-ih-**nef**-rin). These two hormones are given the alternate names of *adrenaline* (ah-**DREN**-uh-**lin**) and *noradrenaline* (**NOR**-ah-**dren**-ah-lin). Epinephrine (adrenaline) is secreted from the "adrenal" (hence the adrenaline name), which is located "upon" (*epi-*) the "kidney" (*nephr*). Norepinephrine (noradrenaline) is "normally" (*nor-*) secreted by the adrenal medulla, along with its close chemical cousin epinephrine (adrenaline), but in much smaller amounts.

Epinephrine circulates widely throughout the bloodstream and generally stimulates the heart to beat faster and harder, and the blood vessels to constrict (narrow), thereby creating a rise in blood pressure. Faster and stronger heart contractions, and a higher blood pressure to push the blood through the vessels more rapidly, help the body under stress to either flight or flee from danger. This danger is usually recognized first within the brain, which then engages the "Fight-or-Flight" Response involving the sympathetic nerves and adrenal hormones. This close functional inter-relationship may thus be called the *sympathetic nerve–adrenal connection*.

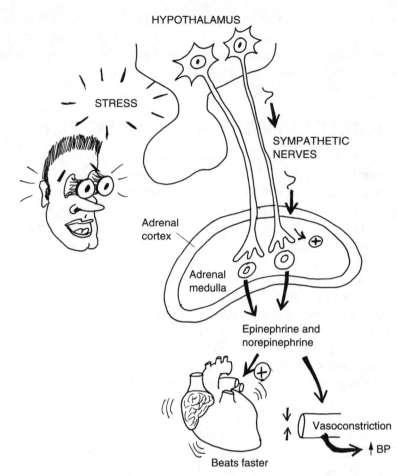

Fig. 15.4 "Fight-or-Flight": The sympathetic nerve–adrenal connection.

Hormones (First Messengers) And Their Partners Inside Cells

We have now traced the operation of several examples of the neuroendocrine system. Within such nerve–gland connections, it is often difficult to tell which part of the connection (the nerve or the gland) comes first in communicating and controlling various aspects of homeostasis. In certain instances, however, the endocrine system, alone, provides the message for initiating specific body responses to *stimuli* (**STIM**-you-**lie**) or "goads."

INSULIN SECRETION AND BLOOD GLUCOSE CONCENTRATION

An important example of such hormone-directed responses is the stimulation of the beta cells of the pancreas by a rise in blood glucose concentration. As Figure 15.5 shows, eating a sugar-rich meal soon results in a rise in blood glucose concentration above the set-point level, and towards the upper limit of its normal range. There is only one major gland (the beta cells of the pancreas) that is strongly excited by this particular stimulus.

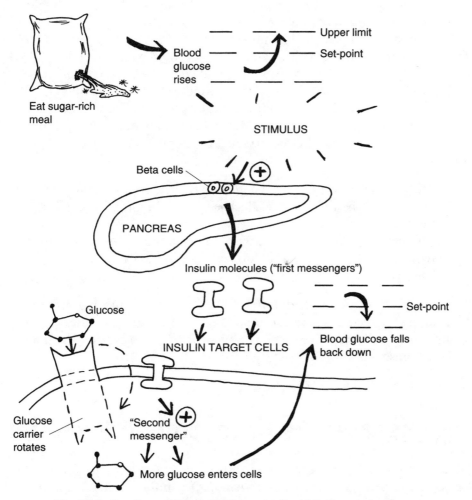

Fig. 15.5 Insulin and "Second Messengers" within its target cells.

The beta cells respond by increasing their rate of secretion of insulin molecules into the bloodstream. Insulin hormone molecules are, then, the "First Messengers" within the bloodstream. By this we mean that the presence of so many insulin molecules in the blood sends a powerful "First Message" to the rest of the body. In words, this chemical message might read: "Hey! Listen up, target cells! The blood glucose concentration has risen near the upper limit of its normal range! Let's do something about it!"

This "First Message" has an effect upon many cells in the body. But the chemical signal is especially strong for the main *insulin target cells: adipocytes (**AH**-dih-poh-sights) or "fat" (adipo-) "cells," hepatocytes (heh-**PAT**-oh-sights) or "liver" (hepat) "cells," and skeletal muscle fibers. When an insulin molecule binds (attaches) to the plasma membrane of one of these target cells, it has essentially delivered the "First Message."*

But the insulin molecule, like most hormone molecules, is far too large to actually enter the target cells. Instead, it operates by activating a "Second Messenger" – one that lies within the target cell. The "Second Messenger" is an intracellular helper molecule. The "Second Messenger" is thus the chemical that actually causes a change in target cell metabolism. In an insulin target cell, for example, "Second Messengers" activated by the binding of insulin to the cell surface help carry out two important changes in cell metabolism:

1. These molecules speed up the *glucose carrier proteins* that act to aid or "facilitate" the diffusion of glucose out of the bloodstream, and into the target cells.
2. Intracellular "Second Messenger" molecules activate a key enzyme required for glycolysis – the utilization of glucose for making ATP energy.

As a result of these two intracellular processes directed by "Second Messengers," glucose leaves the bloodstream in large quantities, enters the insulin target cells, and gets broken down for free energy. Finally, the blood glucose concentration falls back toward its set-point level.

Endocrine Diseases As Uncontrolled Secretion

When endocrine glands secrete their hormones at a "normal" (*normo-*) rate, we say that there is a *normosecretion* (**NOR**-moh-see-**KREE**-shun) of the hormones. Consider, for example, a normosecretion of insulin. In this case, there is just enough insulin secreted to help glucose diffuse out of the blood-stream, and into the target cells, so that a condition of *normoglycemia* (**NOR**-moh-gleye-**SEE**-me-uh) prevails. Normoglycemia is "a blood condition of" (*-emia*) "normal" (*normo-*) "sweetness or glucose" (*glyc*). When the pancreas (and its beta cells) are of a normal size and activity level, then there is a normosecretion of insulin, and a resulting state of normoglycemia within the bloodstream (see Figure 15.6, A).

What happens if a gland *atrophies* (**AH**-troh-**fees**) – shrinks in size – or some of its cells die or are removed? Since just a shrunken mass of epithe-lial cells is left alive and functioning, gland *hyposecretion* (**HIGH**-poh-see-**KREE**-shun) results (see Figure 15.6, B). Hyposecretion is literally a con-dition of "deficient or below normal" (*hypo-*) secretion of some particular hormone.

To illustrate, consider what happens if many of the beta cells become damaged or die, due either to genetic mutations or to an overall atrophy of the pancreas. One serious disease that could result is *diabetes* (**die**-uh-**BEE**-teez) *mellitus* (**MEL**-uh-tus). Diabetes mellitus exactly translates to mean a "honey-sweet" (*mellit*) "passing through" (*diabetes*) of urine from the body. Persons suffering with Diabetes mellitus often have beta cells that do not produce enough normal insulin to significantly reduce the blood glucose level. As a result of such long-term hyposecretion of insulin, too much glu-cose is left behind in the bloodstream.

After glucose builds up over time, a state of *hyperglycemia* (**HIGH**-per-gleye-**SEE**-mee-ah) follows. Hyperglycemia is an "excessive or above nor-mal" (*hyper-*) concentration of glucose within the bloodstream. With so much glucose staying in the blood, eventually some of it is excreted into the urine by the kidneys. The result is technically called *glycosuria* (**gleye**-koh-**SUR**-ee-ah) or "an abnormal condition of" (*-ia*) "glucose" (*glycos*) within the "urine" (*ur*).

Finally, let us ponder the condition of glandular hypertrophy (enlarge-ment) or the presence of gland tumors (Figure 15.6, C). With so many addi-tional epithelial cells being present, the enlarged gland often has a *hypersecretion* (**HIGH**-per-see-**KREE**-shun) of its hormones. This means, of course, that there is an excessive or above normal rate of hormone secretion by the gland. Extreme physiological results often occur.

Let us apply this to the pancreas and its secretion of insulin. When there is a marked enlargement (hypertrophy) of the pancreas and its beta cells, then there is also a marked hypersecretion of insulin. Too much glucose is removed from the bloodstream and transported into the body target cells. With so much glucose gone, severe *hypoglycemia* (**HIGH**-poh-gleye-**SEE**-mee-ah) may occur. [**Study suggestion:** Following the pattern set by the pre-

1, Disorder

2, Disorder

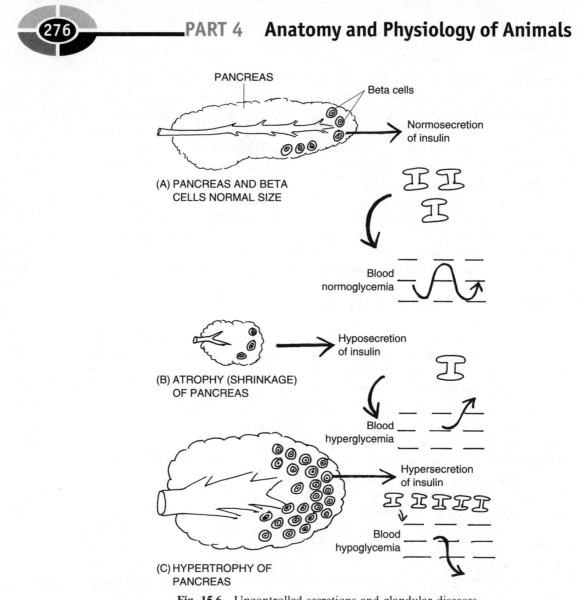

Fig. 15.6 Uncontrolled secretions and glandular diseases.

ceding paragraphs, try to write the literal English translation of hypoglyce-mia on a piece of paper. Check your translation with a friend.]

  Insulin shock or *coma* can result from a severe and prolonged hypoglyce-mia. The neurons in the brain are completely starved of glucose, their major fuel. The person goes into coma or the "deep sleep" of brain failure, and may lose consciousness and die.

3, Disorder

Quiz

Refer to the text in this chapter if necessary. A good score is at least 8 correct answers out of these 10 questions. The answers are listed in the back of this book.

1. A synapse represents:
    (a) A place where the tips of an axon almost touch a muscle fiber
    (b) The major central region of a neuron
    (c) The fluid-filled gap where two neurons nearly come together
    (d) A traveling wave of chemical excitation

2. _____ are masses of epithelial cells specialized for the function of secretion:
    (a) Blood vessels
    (b) Hormones
    (c) Hormone-binding sites
    (d) Glands

3. The *mammary* (**MAH**-mair-ee) *glands* within the female breasts are most accurately classified as:
    (a) Synapses
    (b) Chemical messengers
    (c) Endocrine glands
    (d) Exocrine glands

4. The phrase neuroendocrine system implies that:
    (a) Some parts of the nervous system have no relationship to body secretions
    (b) There are just too darn many body structures and functions to memorize!
    (c) Some nervous structures and endocrine gland structures are functionally related
    (d) Certain glands have unknown body functions that the brain can't understand

5. Antidiuretic hormone is actually produced and secreted by the:
    (a) Posterior pituitary
    (b) Anterior pituitary
    (c) Entire pituitary body
    (d) Hypothalamus

6. TSH exerts a trophic effect upon the _____ gland:
   (a) Testis
   (b) Thyroid
   (c) Thymus
   (d) Tonsil

7. Cortisol generally causes the blood glucose concentration to:
   (a) Fall towards the lower limit of its normal range
   (b) Stay relatively constant at all times
   (c) Soar far beyond the upper limit of its normal range
   (d) Rise towards the upper limit of its normal range

8. Triggers ovulation (release of a mature ovum):
   (a) GH
   (b) ACTH
   (c) FSH
   (d) LH

9. Epinephrine:
   (a) Is exactly the same hormone as noradrenaline
   (b) Typically causes a profound "relaxation response" and deep sleep
   (c) Closely mimics the effects of sympathetic nerve stimulation
   (d) Usually decreases the strength and rate of heart contractions

10. A "Second Messenger" is often required in the endocrine system, since:
    (a) Most hormones don't circulate far enough to reach tissue cells
    (b) Metabolic action does not respond to hormone concentrations
    (c) A lot of hormone molecules are too big to enter their target cells
    (d) An error in interpreting the "First Messenger" frequently occurs

The Giraffe ORDER TABLE for Chapter 15
   (Key Text Facts About Biological Order Within An Organism)

1. _____

2. _____

3. _____

4. _____

5. _____

The Dead Giraffe DISORDER TABLE for Chapter 15
   (Key Text Facts About Biological Disorder Within An Organism)

1. _____

2. _____

3. _____

Blood and the Circulatory System

1, Order

Chapter 15 included an overview of the neuroendocrine system. If you recall, the endocrine glands (such as the thyroid) all secrete hormones directly into the bloodstream. But the bloodstream, however, doesn't carry the blood and its hormones through the body in a straight line. Rather, it carries the blood in a "little circle" (*circul*) (see Figure 16.1).

Chapter 16 thus discusses the *circulatory* (**SIR**-kyuh-luh-**tor**-ee) or *cardio-vascular* (**car**-dee-oh-**VAS**-kyuh-lar) *system*. (This was first mentioned in Chapter 2 and Figure 2.4.) The circulatory (cardiovascular) system is an organ system including the "heart" (*cardi*) and "little vessels" (*vascul*) that carries the blood in a "little circle" that both begins and ends with the heart. Along the way, the blood passes by the main body tissues. In so doing, the blood delivers hormones, glucose, and various other nutrients to the main body tissues. And, as it leaves, the blood picks up a number of waste products from the main body tissues.

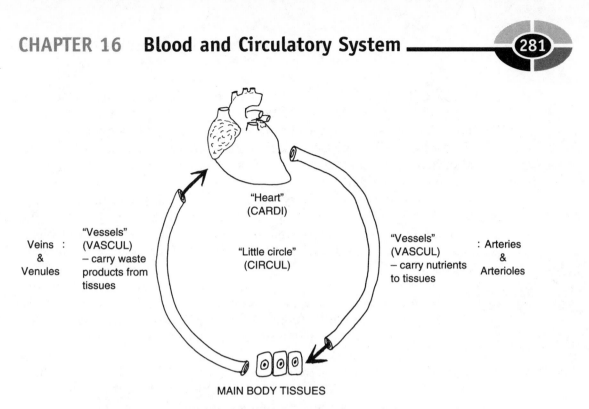

Fig. 16.1   An outline of the circulatory (cardiovascular) system.

The Heart and Blood Vessels: The Pump and Its Pipes

Glance back at Figure 16.1. You will see that if the heart represents a pump for the blood, then the blood vessels represent the pipes carrying it! The *arteries* (**AR**-ter-**ees**) are large-diameter vessels that always carry blood away from the heart. Thus, the arteries immediately receive the blood being pumped from the *ventricles* (**VEN**-trih-**kls**) – the "little belly"-like lower chambers on either side of the heart. Specifically, the *right ventricle* (abbreviated as *RV*) pumps blood out into the *common pulmonary* (**PULL**-mun-**air**-ee) *artery* (abbreviated as *CPA*), which in turn sends blood out towards both "lungs" (*pulmon*). And the *left ventricle* (abbreviated as *LV*) pumps blood out into the *aortic* (ay-**OR**-tik) *arch*. The aortic arch sends the blood from the LV out towards the tissues of the major body systems (other than the lungs) (See Figure 16.2, A).

As the major arteries (such as the common pulmonary artery and aortic arch) travel farther from the heart, they branch into smaller arteries

(Figure 16.2, **B**). And these smaller arteries, in turn, branch into even smaller *arterioles* (ar-**TEER**-ee-ohls) or "little arteries." As the arterioles approach the cells of the body tissues, they branch into the smallest blood vessels of all, the *capillaries* (**CAP**-ih-**lair**-ees). Each capillary is very narrow (much like a strand of hair). This characteristic is reflected in the translation of capillary, which "pertains to a hair."

Since the wall of each capillary is only a single cell thick, nutrients and waste products diffuse across the wall. Nutrients (such as oxygen, O_2, and glucose) diffuse out of the bloodstream, and into the tissue cells. And waste products (such as carbon dioxide, CO_2) diffuse out of the tissue cells, and into the bloodstream.

After the capillaries run past the tissue cells, several of them merge together to form the *venules* (**VEN**-yewls). The venules are the "little veins," in the sense that they connect the capillaries to the much larger *veins*. The veins are all wide-diameter vessels that return blood towards the heart.

Several of the largest veins return blood back to the *atria* (**AY**-tree-ah) – the small "entrance rooms" (*atri*) or chambers located at the top of the heart. Among the biggest sets of veins are the *superior* and *inferior vena* (**VEE**-nah) *cavae* (**KAY**-veye), or "upper and lower cave veins." The superior vena cava (SVC) drains blood down into the *right atrium* (**AY**-tree-um), or *RA*, from the area above the heart, while the inferior vena cava (IVC) returns blood up into the right atrium from the entire region below the heart.

The four *pulmonary veins*, as their name suggests, return blood from the lungs and empty it into the *left atrium, LA*.

THE SYSTEMIC AND PULMONARY CIRCULATIONS

For ease of study, the major vessels and heart chambers are often grouped together into two connected circulations. The *pulmonary* or *right-heart circulation* involves the circulation of blood to, through, and from the lungs. The *systemic* (sis-**TEM**-ik) or *left-heart circulation*, in contrast, represents the circulation of blood to, through, and from the tissues of all the major body organ systems (except for the lungs). [**Study suggestion:** Review Figure 16.2. From this diagram, which specific chamber of the heart begins the pulmonary circulation? Which specific blood vessels end the pulmonary circulation? What particular heart chamber begins the systemic circulation? What particular blood vessels end the systemic circulation?]

2, Order

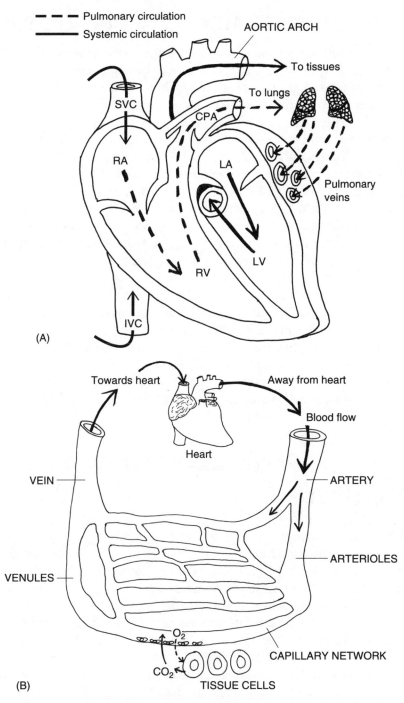

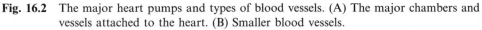

Fig. 16.2 The major heart pumps and types of blood vessels. (A) The major chambers and vessels attached to the heart. (B) Smaller blood vessels.

COMPARISON TO CIRCULATIONS IN OTHER VERTEBRATES

1, Web

The four-chambered human heart (two upper atria, two lower ventricles) plus a *double circulation* (pulmonary circulation plus systemic circulation) is pretty typical of the heart and circulation found in both birds and other mammals. Being endotherms (or homeotherms) that regulate their own internal body temperature (Chapter 12), birds and mammals have a great need for energy and oxygen to keep their body heat always at a fairly high level. A double circulation plus powerful four-chambered heart provide both the required force to pump a large volume of heated blood and a separate pulmonary circulation that adds plenty of oxygen to the blood.

The pulmonary (right-heart) circulation in mammals, therefore, is specialized for adding oxygen to the bloodstream. But the systemic (left-heart) circulation specializes in extracting oxygen from the bloodstream and delivering it to the tissues of the major organ systems. In the process, carbon dioxide is picked up from the tissues.

The circulation in vertebrates that are heterotherms (poikilotherms), such as fish, amphibians, and reptiles, is quite another matter. Fish have a two-chambered heart with only a single circulation. Amphibians rely upon a three-chambered heart (two atria, one ventricle) beating within their chests. Reptiles, likewise, possess a three-chambered heart, with two atria and one ventricle that is partially subdivided. Both amphibians and reptiles have double circulations, but there is much less specialization of function of the two circulations.

Internal Anatomy, Pacemaker Tissue, and Valves of the Heart

The powerful four-chambered heart of humans and other mammals has a complex internal anatomy (Figure 16.3). Consider, for example, the *myocardium* (**my**-oh-**CAR-dee**-um) or "heart" (*cardi*) "muscle" (*my*). The cardiac muscle fibers of the myocardium are arranged in a fairly circular pattern around the heart, so when they contract, they squeeze the blood out of the heart chambers like a noose tightening around a bag. The blood from the atria is pushed down into the ventricles, through a pair of one-way valves. Not surprisingly, these valves are called the *right* and *left atrioventricular* (**ay-**tree-oh-ven-**TRIK**-you-lar) *valves*. These two *A-V* (atrioventricular) valves essentially act as one-way doors. They are actually flaps of connective tissue that are pushed open from above by blood in the atria.

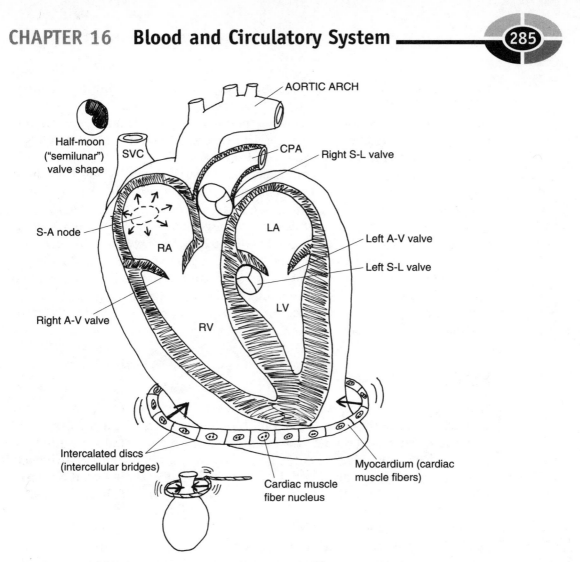

Fig. 16.3 Some internal structures and functions of the heart.

PACEMAKER TISSUE

The atria contract because they are excited by *cardiac pacemaker cells.* These pacemaker cells are actually modified cardiac muscle fibers that are self-exciting. Sodium (Na^+) and other charged particles are automatically let into the pacemaker cells at a certain rate or rhythm. This happens because the proteins in the membranes of the pacemaker cells tend to shift around, allowing ions to enter and excite the pacemaker cells or "turn themselves on."

The main cardiac pacemaker area is called the *sinoatrial* (**sigh-no-AY**-tree-al) or *S-A node* (**NOAD**). This region is called a node because it is somewhat rounded, like a "knot." The word sinoatrial indicates its anatomical location. The S-A node lies in the outer wall of the right "atrium," just below the entrance of the superior vena cava – a major *venous* (**VEE**-nus) *sinus* (**SIGH**-nus). (A sinus, in general, is a large vein that is shaped like a "hollow bay" [*sin*], holding a considerable volume of blood.)

As the *primary cardiac* pacemaker, the sinoatrial node excites itself first. The excitation then spreads from one cardiac muscle fiber to another, due to the presence of *intercalated* (**in**-ter-**KAY**-lay-ted) *discs* between them. These intercalated discs appear as thick dark lines when viewed through a compound microscope. They are actually *intercellular bridges* "inserted between" (*intercalat*) adjacent cardiac muscle fibers. Thus, when the S-A node fires off an action potential, this traveling wave of excitation moves from one cardiac muscle fiber to another, across the intercalated discs connecting them end-to-end. Eventually, both atria become excited and contract. And finally, both ventricles become excited.

SEMILUNAR VALVES

After the ventricles are excited, their walls contract. The blood in the right ventricle is pushed up into the common pulmonary artery through the *right semilunar* (**sem**-eye-**LOO**-nar) *valve*. Likewise, the blood contained within the left ventricle is pumped up into the aortic arch through the *left semilunar valve*. Both of these valves get their names from the shapes of their flaps, which look like "partial or half" (*semi-*) "moons" (*lun*). This is reflected in their abbreviation, *S-L*.

Note back in Figure 16.3 that the myocardium is thicker in the walls of the ventricles, compared to the atria. And observe that the myocardium is thickest of all in the wall of the left ventricle. [**Study suggestion:** Ask yourself, "Why is the myocardium thicker in the wall of the left ventricle, compared to that of the right ventricle?" Hint: What happens to a skeletal muscle when you repeatedly exercise it?]

The Cardiac Cycle

After the chambers of the heart are excited, they go into a state of *systole* (**SIS**-toh-**lee**) or "contraction" and emptying. It is during systole that the blood is pumped from each of the ventricles, and out into their major arteries. After systole, comes *diastole* (die-**AH**-stoh-**lee**) or "relaxation" and filling. While the atria are still contracting in systole, the ventricles below them are in a state of diastole, so that they receive blood through the right and left atrioventricular valves.

The *Cardiac Cycle* is one heart beat or one complete cycle of contraction (systole) plus relaxation (diastole) of all four chambers of the heart. There are also at least two *heart sounds* that can be heard through a stethoscope placed onto the chest or back. The *1st heart sound* is commonly described as "lubb." It is due to closure of both A-V valves, which occurs when the ventricles start going into systole. The ventricles push the A-V valve flaps shut from below. As a result, a back-flow of blood from the ventricles, and back up into the atria, is usually prevented.

3, Order

The *2nd heart sound* is usually represented as "dupp." This sound occurs because of the closing of both semilunar valves at the beginning of ventricular diastole. The blood in the common pulmonary artery/aortic arch above each ventricle starts to fall back downward, due to the force of gravity. As the blood falls down, it catches the semilunar valve flaps and slams them shut, thereby preventing a back-leak into the ventricles. [**Study suggestion:** Glance back at Fig. 16.3, to help you visualize what's happening during the two heart sounds.]

HEART MURMURS

Sometimes the heart valve flaps do not fit tightly together. Consider what may happen after *bacterial endocarditis* (**en**-doh-car-**DIE**-tis). This disease involves an "inflammation of" (*-itis*) the "inner" (*endo-*) lining of the "heart" (*cardi*), due to infection with bacteria. During the process of inflammation, the flaps of certain valves (especially those of the left A-V valve) become swollen. When the inflammation finally heals, the valve flaps may become pulled back and distorted.

1, Disorder

Thus, there is a *turbulent* (**TUR**-byuh-lunt) back-flow of blood – a back-flow which is highly disorderly and in a state of "turmoil"– through the distorted and ill-fitting heart valve flaps. The resulting noisy, abnormal heart sounds are called a *heart murmur*.

Blood Pressure and Blood Flow

In general, *blood pressure* (BP) is a pushing force exerted against the blood and against the walls of the blood vessels (see Figure 16.4). The blood pressure (BP) is at its highest in the major arteries attached directly to the heart

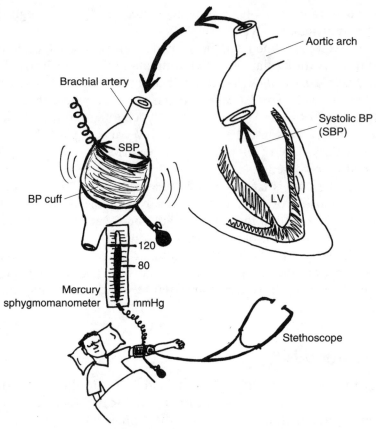

Fig. 16.4 Blood pressure (BP), blood flow, and the brachial artery.

(such as the aortic arch and common pulmonary artery). The BP then progressively decreases with greater distance from the heart.

4, Order

Blood pressure is critically important because it is the main force causing blood flow. There is a *blood pressure gradient* (**GRAY**-dee-unt), a series of downward "steps" from a region of higher to a region of lower blood pressure. Within the systemic circulation, for example, the BP is highest within the aortic arch. As the aortic arch progressively branches into smaller and smaller arteries farther away from the heart, the BP also declines. This creates a blood pressure gradient or series of downward steps in blood pressure in vessels farther from the heart. Hence, the blood is pushed down a BP gradient, so that it flows into lower-pressure arteries in a series of steps.

One of these farther arteries is called the *brachial* (**BRAY**-kee-al) *artery*, which is named for its location within the upper "arm" (*brachi*). The brachial artery is the most frequent site used for taking someone's blood pressure, utilizing a stethoscope and *sphygmomanometer* (**sfig**-moh-muh-**NAHM**-uh-ter). A sphygmomanometer is literally "an instrument used to measure" (*-meter*) the "throbbing pulse" (*sphygmo*) at certain "intervals" (*mano*).

To be sure, one does hear a dull throbbing sensation when the bell of a stethoscope is placed over the brachial artery and the arm cuff of a sphyg-momanometer device is inflated around it. The sphygmomanometer is usually marked off in units of *millimeters of mercury*, abbreviated as *mmHg* (Hg is the chemical symbol for the element, mercury).

As the air is slowly let out of the inflated arm cuff, the first dull throbbing noise one hears through the stethoscope is called the *systolic* (**sis-TAHL**-ik) *blood pressure*, or *SBP*. The systolic blood pressure (SBP) is the pressure created by the systole (contracting and emptying phase) of the left ventricle of the heart. A slug of blood is pushed out of the left ventricle with its contraction. Flowing progressively down a BP gradient, this slug of blood finally enters the brachial artery. The blood bulges out the artery somewhat as it passes through it, thereby creating a thumping sound. For a resting adult, the systolic blood pressure (SBP) is usually recorded as about 120 mmHg.

The *diastolic* (**DIE**-ah-stahl-ik) *blood pressure*, or *DBP*, is the blood pressure associated with the diastole (relaxing and filling) phase of the left ventricle. Upon thoughtful reflection, you might well ask, "Why isn't the diastolic BP just 0 mmHg? Isn't diastole the resting and filling phase, when the ventricle isn't even contracting or creating any blood pressure? So, why should there even be any diastolic BP, at all?"

Good question! Glance back at Figure 16.4. Note that the brachial artery bulges out with the force of the systolic BP against its walls. Now, when the ventricle stops contracting, and diastole begins, there is a powerful elastic recoil, or snapping back force, created by the stretched brachial artery wall coming back to its non-stretched shape. This force of elastic recoil or snapping back of the stretched brachial artery wall is what creates the diastolic BP. The diastolic blood pressure is, then, in a sense, a residual or left-over blood pressure, the force of blood pressure temporarily "stored" in the bulged-out walls of the brachial artery during systole. Because it is a residual BP, the diastolic BP is normally significantly lower than the systolic BP. It averages about 80 mmHg in a resting adult.

The stethoscope and sphygmomanometer measure a resting adult's blood pressure as about 120/80 mmHg. The SBP is 120, the DBP is 80. The diastolic BP is usually recorded as the level of blood pressure where the dull thumping sound heard through the stethoscope just disappears. [**Study suggestion:** Ask yourself, "Why does the thumping sound disappear just below the recorded level of the diastolic BP?" **Hint:** Think about what is happening to the wall of the brachial artery.]

Hypertension, Hypotension, and Arteriosclerosis

We have been talking about the normal blood pressure, which has an average or set-point level of about 120/80 mmHg. The normal BP can rise to an upper limit of about 140/90 mmHg. And it can fall down to a lower limit of about 100/60 mmHg. The distance between these upper and lower normal readings creates the *normal range for blood pressure*, a condition that can technically be called *normotension* (**NOR**-moh-**TEN**-shun). We use this term because blood "pressure" represents the amount of *tension* exerted against the blood. Figure 16.5 (A) illustrates a state of normotension: that is, a state of relative constancy or homeostasis of blood pressure within its normal range.

Unfortunately, however, blood pressure doesn't always remain within its normal range. Say that a lumberjack accidentally cuts his brachial artery with a chainsaw while felling a large oak tree. The blood spurts out in hot red jets, resulting in a severe and possibly fatal *hemorrhage* (**HEM**-uh-rij) – a "bursting out" (*-orrhage*) of "blood" (*hem*). When so much blood is lost, there isn't much blood left to press against the arterial wall. Thus, blood pressure steeply declines. It may even reach a state of *hypotension* (**HIGH**-poh-**TEN**-shun). Hypotension is a condition of "below normal or deficient" (*hypo-*) blood "pressure" (*tens*). Specifically, hypotension is a blood pressure significantly less than 100/60 mmHg (Figure 16.5, B).

2, Disorder

3, Disorder

At the opposite extreme lies *hypertension* (**HIGH**-per-**TEN**-shun) – an "excessive or above normal" (*hyper-*) blood pressure. Hypertension is a blood pressure significantly above the upper normal limit of approximately 140/90 mmHg (Figure 16.5, C).

"So what if a person has hypotension or hypertension?" a skeptic might inquire. With hypotension, the person may easily faint. And if the condition is severe, there may be *circulatory shock* or *coma*, due to a lack of blood pressure pushing blood up to feed the brain. In the case of hypertension, the chronically above-normal pressure may overstretch and thin out the walls of arteries, creating *aneurysms* (**AN**-yuh-**riz**-ums). Aneurysms are abnormally "widened up" arteries, which are highly susceptible to being ruptured. And when an aneurysm ruptures, there may be a large amount of internal bleeding. Persons suffering a stroke, or *cerebrovascular* (seh-**REE**-broh-**VAS**-kyoo-lar) *accident*, for instance, may well have experienced a ruptured aneurysm of their cerebral blood vessels covering the brain. Whatever particular body functions the oxygen-and-blood-deprived brain area carried out, are then partially or totally lost.

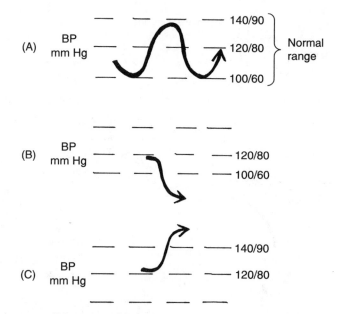

Fig. 16.5   The three possible states of blood pressure (BP). (A) Normotension. (B) Hypotension. (C) Hypertension.

ARTERIOSCLEROSIS

Closely related to hypertension is the condition called *arteriosclerosis* (ar-**teer**-ee-oh-sklair-**OH**-sis). Arteriosclerosis is "an abnormal condition of" (*-osis*) "hardening" (*scler*) of the walls of an "artery" (*arteri*).

There is a *positive feedback* relationship between arteriosclerosis and hypertension. You may remember (Chapter 14) that a negative feedback system removes or corrects a particular change. It thus tends to re-establish or maintain a state of Biological Order.

Positive feedback, however, is a system that magnifies or accelerates change in some particular direction. Whatever is happening already, positive feedback makes it even more extreme, or worse. Thus, positive feedback is sometimes referred to as a *vicious cycle*. It magnifies or accelerates changes that often veer out of control, thereby promoting Biological Disorder.

Consider hypertension and arteriosclerosis (Figure 16.6). When the BP is consistently elevated above its normal range, then it tends to damage the elastic fibers within the arterial walls. [**Study suggestion:** Picture a garden hose

4, Disorder

292

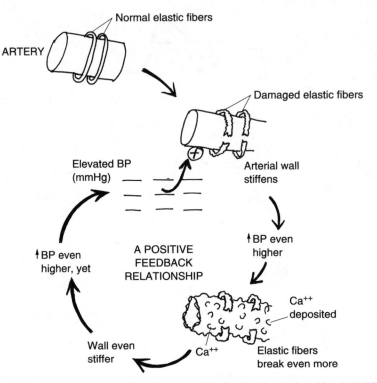

Fig. 16.6 Hypertension and arteriosclerosis: A positive feedback relationship. (*NOTE*: Blood pressure is abbreviated as BP.)

with an adjustable nozzle at the end. It is attached to a faucet that is left turned on all summer, with the nozzle closed. What will eventually happen to the garden hose?] When the elastic fibers are damaged, the arterial wall gets stiffer, and when the arterial wall gets stiffer, the BP goes up even more. This makes the hypertension worse. And as the hypertension worsens, the elevated pressure damages the elastic fibers to an even greater extent.

Hence, there is a positive feedback relationship (vicious cycle) between hypertension and arteriosclerosis. The worse the hypertension gets, the more the arterial wall hardens, and the more the arterial wall hardens, the worse the hypertension gets. Adding to the damage, calcium ions are pushed out of the bloodstream by the high BP, becoming embedded in the arterial wall. So the once soft, pliable, garden hose-like normal artery progressively transforms into a rock-hard, stiff, *calcified* (**KAL**-sih-**feyed**) tube that more resembles a heavily limed-up lead water pipe!

The Blood Connective Tissue

Having discussed the heart as a pump and the blood vessels as tubes for carrying flowing blood, it is now time to consider the blood itself. Blood is a red, sticky, connective tissue that occupies about 5–6 *liters* (**LEE**-ters) of volume in an average-sized adult.

PLASMA

At first, it may seem odd to classify blood as a connective tissue. After all, it contains no connective tissue fibers (such as collagen or elastic fibers) that directly strap body parts together. Nevertheless, blood is considered a *special connective tissue*, because it contains *plasma* (**PLAZ**-muh). Plasma is the clear, watery, intercellular substance found "between" (*inter-*) the "cells" (*cellul*) of the bloodstream. Because it circulates throughout the body within the bloodstream, plasma acts like a functional connective tissue rather than a structural one. Recall that hormones (Chapter 15), for instance, are First Messengers secreted by the endocrine glands and carried throughout the body within the bloodstream. It is the plasma, not the blood cells per se, that does this functional connecting or carrying of the hormones towards their various target cells.

5, Order

The plasma is basically a *saline* (**SAY**-leen) or "salt"-containing solution consisting of 0.9% NaCl (sodium chloride) solute (Chapter 4) dissolved in water solvent. In addition to carrying hormones, nutrients such as glucose and O_2, and waste products like CO_2, the blood also contains a number of important *plasma proteins*. Some of these plasma proteins are critical for blood clotting. Still others play a role in such processes as body defense from foreign invaders.

FORMED ELEMENTS IN THE BLOOD

Besides the plasma, there are also many *formed elements* within the bloodstream. Formed elements are cells and cell fragments within the blood that have a definite shape or form.

When a sample of human blood is stained and viewed under a compound microscope (Figure 16.7), several major types of formed elements can be identified. Most numerous among these are the *erythrocytes* (air-**RITH**-roh-**sights**) or "red" (*erythr*) blood "cells" (*cytes*). Erythrocytes (red blood cells) are *anucleate* (**ay**-**NEW**-klee-aht); that is, they are "without" (*a-*) any

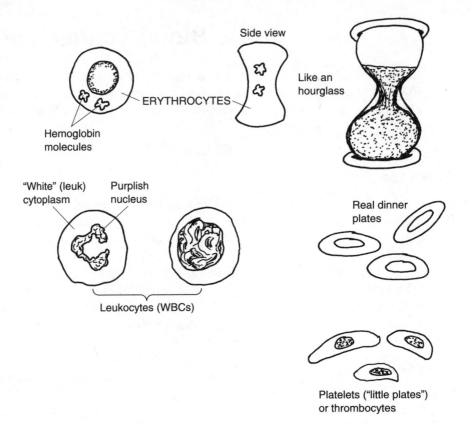

Fig. 16.7   The major formed elements of the blood.

nucleus. They are also quite special in that they are shaped like _biconcave_ (buy-**KAHN**-cave) _discs_, being "caved-in" on "both" (_bi-_) sides. Viewed from the side, this makes them look like red hourglasses! The red color is mainly due to the presence of _hemoglobin_ (**HEE**-moh-**glohb**-in). Hemoglobin is a reddish-colored, "globe"-shaped (_glob_), "protein substance" (_-in_) found within the cytoplasm of the red "blood" (_hem_) cells. There may be as many as 10,000 hemoglobin molecules present within a single erythrocyte. The main job of these thousands of hemoglobin molecules is carrying oxygen (O_2) molecules through the bloodstream, and to the tissue cells.

Leukocytes (**LEW**-koh-**sights**) are the "white" (_leuk_) blood "cells" (_cytes_). These cells typically have a large, purplish-staining nucleus, but they are named for the clear, whitish appearance of their cytoplasm. (This is in marked contrast, of course, to the red, hemoglobin-rich cytoplasm of the erythrocytes.) The leukocytes have significant roles in protecting the body from various foreign invaders. (Details are given in Chapter 17.)

The third common formed element are the *platelets* (**PLAY**-teh-**lets**) or *thrombocytes* (**THRAHM**-buh-**sights**). The platelet name comes from their shape – "little plate"-like fragments of disintegrated bone marrow cells. The thrombocyte name derives from their function – "clotting" (*thromb*) "cells." About 1/3 the size of an erythrocyte, the platelets (thrombocytes) are scattered here and there in small groups, throughout the plasma. These purplish-colored, plate-like cell fragments have very sticky surfaces. Thus, whenever a blood vessel ruptures and hemorrhages, the platelets soon collect around the open hole and stick to a network of thin, "fiber"-like strands, composed of the protein, *fibrin* (**FEYE**-brin). As the platelets collect and stick to one another, a "clot" or *thrombus* (**THRAHM**-bus) is soon formed. The hole is blocked, hemorrhaging stops, and wound healing follows.

Serum Cholesterol, Atherosclerosis, and "Heart Attacks"

When a person has a so-called "heart attack," it usually involves a problem with abnormal clotting or an *occlusion* (uh-**KLEW**-zhun) – "shutting up" – of the blood vessels serving the myocardium. The site of the troublesome occlusion (shutting up) is somewhere within the *coronary* (**KOR**-uh-**nair**-ee) *arteries*. These vessels are named for their resemblance to a "crown" (*coron*) encircling the top of the heart, just below the *auricles* (**OR**-ih-**kls**) or "little ear"-like flaps. The right and left coronary arteries look like a red-colored crown slipped down over the ears of a real prince (Figure 16.8, A)!

CHOLESTEROL AND ATHEROSCLEROSIS

The blood *serum* (**SEER**-um) is the clear, yellowish watery portion of the blood plasma that remains after a clot (thrombus) has been formed. The blood serum contains *cholesterol* (kah-**LES**-ter-ahl). Cholesterol is a white, fatty substance that tends to be laid down as *atheromas* (**ah**-ther-**OH**-mahs) or "fatty" (*ather*) "tumors" (*-omas*) whenever its concentration becomes too high.

Atherosclerosis (**ah**-ther-oh-slair-**OH**-sis) is a fatty hardening of the arteries that occurs when excessive blood cholesterol is deposited as atheromas (fatty tumors) upon the inner walls of arteries. Atherosclerosis is especially harmful in cases of *coronary artery disease*, where atheromas can build up and suddenly occlude (stop up) a coronary artery opening (Figure 16.8, B).

5, Disorder

(A) THE CORONARY ("pertaining to a crown") ARTERIES

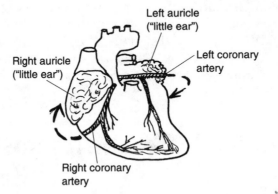

Left auricle ("little ear")

Right auricle ("little ear")

Left coronary artery

Right coronary artery

(B) ATHEROSCLEROSIS AND CORONARY HEART DISEASE

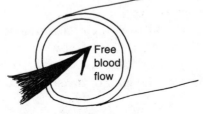

Free blood flow

Normal coronary artery

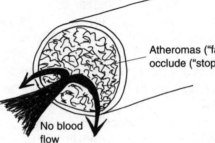

Atheromas ("fatty tumors") occlude ("stop up") the vessel

No blood flow

Severe atherosclerosis of coronary artery

Fig. 16.8   Atherosclerosis and occlusion of the coronary arteries.

As a result of this occlusion, the myocardium in the heart wall is suddenly choked off from oxygen, glucose, and other nutrients previously delivered by the coronary artery. The stricken person may experience crushing chest pain or *angina* (an-**JEYE**-nuh) *pectoris* (**PEK**-tor-is). A *coronary bypass operation* may be performed. In this operation, the surgeon bypasses the occluded parts of the coronary arteries by implanting small sections of veins obtained from other areas of the body. Hopefully, the operation works, and fresh blood is shunted past and around the blocked vessel areas to successfully feed the nutrient-starved cardiac muscle.

Quiz

Refer to the text in this chapter if necessary. A good score is at least 8 correct answers out of these 10 questions. The answers are listed in the back of this book.

1. The circulatory system is literally named for its characteristic of:
    (a) Transporting nutrients to the body tissues
    (b) Carrying waste products from the body tissues
    (c) Traveling through the body in a straight line
    (d) Tracing a circle in its journey to and from the heart

2. The smallest branches of the arteries:
    (a) Capillaries
    (b) Veins
    (c) Venules
    (d) Arterioles

3. The _____ are the "entrance rooms" at the top of the heart:
    (a) Atria
    (b) Auricles
    (c) Ventricles
    (d) Myocardia

4. The left-heart circulation is alternately called the _____ circulation:
    (a) Pulmonary
    (b) Systemic
    (c) Cardiorespiratory
    (d) Cerebrovascular

5. The powerful cardiac muscle portion of the heart wall:
    (a) Semilunar valves
    (b) Endocardium
    (c) Myocardium
    (d) Pericardium

6. The primary pacemaker of the heart:
    (a) Myocardium
    (b) Left atrium
    (c) Sinoatrial node
    (d) Right A-V valve

7. Half-moon shaped flaps that control the entry of blood into the aortic arch and common pulmonary artery:
    (a) Atrioventricular valves
    (b) Z-lines
    (c) Semilunar valves
    (d) Crescent muscles

8. The relaxation and filling phase of each heart chamber:
    (a) Diastole
    (b) Fibrillation
    (c) Systole
    (d) Bacterial endocarditis

9. Heart murmurs represent:
    (a) Turbulent back-flow of blood through leaky valve flaps
    (b) Smooth and efficient blood flow through healthy open valves
    (c) Extensive arteriosclerosis
    (d) A false appearance of atherosclerosis

10. The vessel most frequently used to take a person's blood pressure:
    (a) Superior vena cava
    (b) Brachial vein
    (c) Common pulmonary artery
    (d) Brachial artery

The Giraffe ORDER TABLE for Chapter 16
(Key Text Facts About Biological Order Within An Organism)

1. _____
2. _____
3. _____
4. _____
5. _____

The Dead Giraffe DISORDER TABLE for Chapter 16
(Key Text Facts About Biological Disorder Within An Organism)

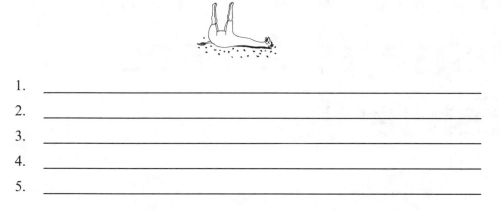

1. _____
2. _____
3. _____
4. _____
5. _____

The Spider Web ORDER TABLE for Chapter 16
(Key Text Facts About Biological Order Beyond the Individual Organism)

1. _____

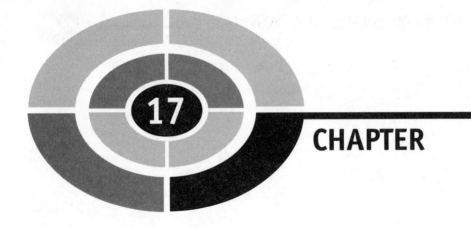

CHAPTER

Immune and Lymphatic Systems: "The Best Survival Offense Is A Good Defense!"

Chapter 16 featured the blood circulatory (cardiovascular) system and briefly mentioned the role of certain plasma proteins in helping to protect the body from foreign invaders. The most appropriate term to describe this protection is *immunity* (ih-**MYEW**-nih-tee) – "a condition of" (*-ity*) "not serving" (*immun*) disease. Thus, when a living organism has an immunity, or it is *immune* (ih-**MYEWN**), then it does not serve or buckle-under to the onslaught of a particular foreign invader or disease. Such immunity is accomplished by the operation of a strong body defense. (Recall from Chapter 6 that immune also means "safety.")

The Lymphatic System: A Shadow Circulation of the Blood

Closely associated with immunity is the *lymphatic* (lim-**FAT**-ik) *system*. The lymphatic system consists of a collection of *lymphatic organs* and *lymphatic vessels* running between them. In a practical sense, the lymphatic system can be thought of as a shadow circulation of the blood circulation. The main reason is that the *lymphatic capillaries* run side-by-side (like a shadow) along the tiny blood capillaries (Figure 17.1).

1, Order

The lymphatic capillaries are special in that they are dead-ended. Since they are closed at the far end, their fluid contents, the *lymph* (**LIMPF**), flows in one direction only – towards the heart. The word, lymph, literally means "clear spring water." This meaning derives from the usually clear, watery appearance of the lymph. The lymph is actually a *filtrate* (**FILL**-trait) – filtration product – of the blood, the blood in the nearby capillaries. The blood pressure pushes outward, thereby filtering water, NaCl, and various foreign objects (such as dirt, bacteria, or cancer cells) out of the blood and into the lymphatic capillaries. Since the lymph usually contains no erythrocytes, it looks clear, rather than red-colored. But eventually, the lymphatic

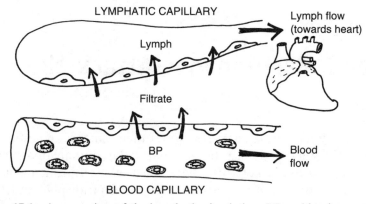

Fig. 17.1 An overview of the lymphatic circulation. BP = blood pressure.

vessels drain their cleansed lymph into several major blood veins that flow into the top of the heart.

THE RETICULOENDOTHELIAL SYSTEM

2, Order

The lymphatic system is sometimes given the alternate name of *reticuloendothelial* (reh-**TIK**-you-loh-**en**-do-**THEE**-lee-al) *system*. The reticuloendothelial or *R-E* system is a "little network" (*reticul*) of lymphatic vessels that are lined by *endothelial* (**en**-doh-**THEE**-lee-al) *cells*. Endothelial cells are flat, scale-like cells that line both blood vessels and lymphatic vessels. And because these endothelial cells are so flat, lymph (and its contained dirt, bacteria, or cancer cells) are readily filtered through or between them.

Since the lymphatic (reticuloendothelial) system receives dirt, bacteria, debris, viruses, and cancer cells that have been filtered out of the bloodstream, it serves a critical role in immunity (body defense) by cleaning up the blood, and then returning the cleansed fluid back to the blood. Remember this general saying: "From the blood the lymph is formed, and back to the blood the lymph doth return." The first lymph filtered from the bloodstream is dirty (in the sense that it often carries contaminants or foreign invaders), while the final lymph is clean (in the sense that the contaminants and foreign invaders have been removed). Therefore, in the process of filtering and cleaning the material that is leaked out of the blood capillaries, the lymphatic (reticuloendothelial) system provides immunity.

3, Order

We can summarize the important inter-relationships by this equation:

$$\text{LYMPHATIC SYSTEM} = \text{RETICULOENDOTHELIAL (R-E)}$$
$$\text{SYSTEM} = \text{IMMUNE SYSTEM}$$

Antibodies and Macrophages Attack the Antigens

Most cells carry chemical markers upon their surface membranes that uniquely identify them. Thus, cells in a particular human body have their own chemical surface markers that mark them as "self." Such cells are not attacked by the body's immune system. Foreign cells transplanted from some other human body, or from a bacterium or cancer cell, however, have different surface markers that mark them as "non-self." The general name for such surface markers is *antigens* (**AN**-tih-**jens**).

The word, antigen, means "produced" (*-gen*) "against" (*anti-*). This meaning reflects the fact that antigens are "non-self" marker proteins that label particular cells as foreign. Therefore, the antigens are foreign proteins that cause *antibodies* (**AN**-tih-**bah**-dees) to be "produced against" them. Antibodies, in turn, are proteins produced by the body's immune system that attack and destroy foreign antigens. The overall process is called an *antigen–antibody reaction.*

The antigen–antibody reaction comes about after a series of preceding steps (Figure 17.2). The first step is identification of a foreign cell and its surface antigen by a *thymic* (**THIGH**-mik) *lymphocyte* (**LIMPF**-oh-**sight**) within the body tissues. The thymic lymphocyte is also abbreviated as *T-lymphocyte* (or as a T-cell, mentioned back in Chapter 6). The T-lymphocytes prowl around within the extensive network of the reticuloendothelial system and act much like scouts. They send out a chemical signal whenever a foreign antigen is encountered.

The *bone marrow* or *B-lymphocytes* receive the chemical messages from the T-lymphocyte scouts. The B-lymphocytes then undergo a marked differentiation (process of becoming specialized or different). They transform into *plasma cells.* Plasma cells have a prominent "clock face" nucleus when viewed through a compound light microscope. There is dark *chromatin* (kroh-**MAT**-in) visible – strands of DNA that have not yet coiled together to create chromosomes. These chromatin fragments are arranged in a circular fashion around the edges of the nucleus, giving it a distinct "clock face" appearance.

It is the plasma cells that actually produce the antibodies. Once produced, the individual antibody molecules attach to the foreign antigens, like two pieces of a jigsaw puzzle fitted together. The result we have called an antigen–antibody reaction. When the antibody combines, it causes a lysis (breakdown) of the invading cell carrying the foreign antigen. Thus, millions of invading or abnormal cells (and their antigens) are efficiently ruptured and scattered into tiny pieces.

Moving nearby is a defensive army of *wandering macrophages* (**MAH**-kroh-**fah**-jes), or "large" (*macr*) "eating" (*phag*) cells. Such wandering macrophages often include the *monocytes* (**MAHN**-oh-**sights**). The monocytes are a type of leukocyte that has a "single" (*mono-*), large, horseshoe-shaped nucleus. The monocyte can creep out of the blood in capillaries by making amoeboid (amoeba-like) movements through their extremely thin walls. They enter the surrounding tissue and extend their cytoplasm like twin arms or pseudopodia. The monocytes readily surround and engulf

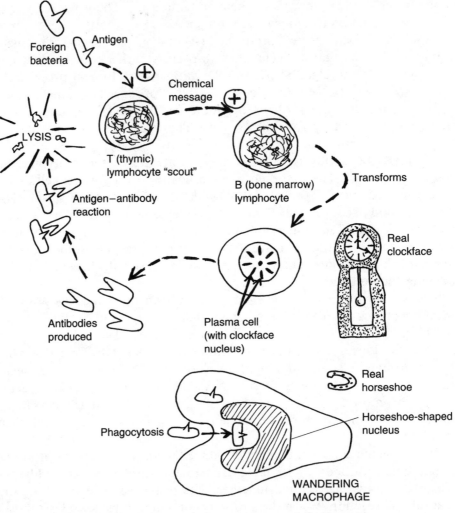

Fig. 17.2 Antibodies and macrophages attack foreign antigens.

entire invading cells and their foreign antigens by means of phagocytosis (cell eating).

Hence, by two major processes, antigen–antibody reactions and phagocytosis, a state of immunity from disease is usually accomplished within the human body.

4, Order

The Major Lymphatic Organs

So far, we have mainly been focusing upon the lymphatic vessels and the defensive reactions of the immune system. But, remember that the lymphatic system consists of a collection of lymphatic organs, as well as lymphatic vessels.

THE LYMPH NODES

The most widespread lymphatic organs are the *lymph nodes* (**NOADS**). The lymph nodes are a group of small, bean-like organs scattered in clusters in various parts of the body (Figure 17.3). *Afferent* (**AF**-fer-**ent**) *lymphatic vessels* "carry" (*fer*) dirty lymph "towards" (*af-*) the lymph nodes. *Efferent* (**EE**-fer-**ent**) *lymphatic vessels* carry clean lymph "away from" (*ef-*) them.

THE THYMUS GLAND

A most unusual lymphatic organ is the *thymus* (**THIGH**-mus) *gland*. The word, thymus, comes from the Ancient Greek for "warty outgrowth," reflecting the bumpy appearance of this endocrine gland. The thymus is a thin, flat gland lying just deep to the *sternum* (**STER**-num) or "breastplate." This gland consists of two bumpy-looking lobes. The thymus gland secretes the hormone, *thymosin* (thigh-**MOH**-sin), which stimulates the activity of the lymphocytes and other parts of the body's immune system. It also produces *thymic* (**THIGH**-mik) *lymphocytes* (introduced as T-cells or T-lymphocytes, earlier).

The thymus gland is most prominent in young humans and other mammals. (In lambs, it is called the "throat sweetbread," because it is often eaten as sweet-tasting meat.) It reaches its maximum size at puberty, then progressively decreases in size. In most adults, the thymus is completely gone, having been replaced by fatty connective tissue. The thymus is thought to play an important role in the development of *immune competence* in youngsters – a growing ability to ward off various diseases. The disappearance of the thymus in adults may be related to the gradual decline of immune competence seen in older persons, thereby making them more susceptible to cancer and pneumonia.

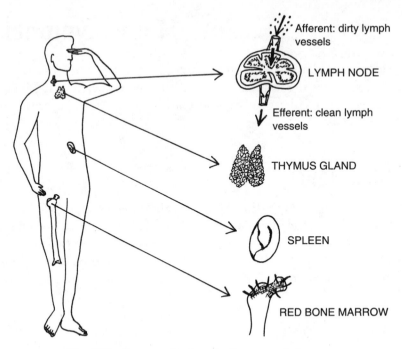

Fig. 17.3 Some major lymphatic organs in humans.

RED BONE MARROW

The red bone marrow found within spongy bone (Chapter 13) is a third important type of lymphatic organ. Besides its role in hematopoiesis (blood cell formation), the red bone marrow is a source of B-lymphocytes. As you may recall, the B-lymphocytes differentiate into plasma cells when they are chemically signaled by the T-lymphocytes that a foreign invader (antigen) is present. And these plasma cells, in turn, produce antibodies.

SPLEEN

In humans, the *spleen* is a dark red organ attached to the left side of the stomach. It looks somewhat like a thick, crescent-shaped-roll (croissant), and it rather feels like one, being soft and spongy to the touch. Besides hematopoiesis, the spleen (like the red bone marrow) is involved in the recycling and destruction of old, beaten-up erythrocytes. It also stores blood and contains a lot of lymphatic tissue. The spleen is rich in lymphocytes, plasma cells, and wandering macrophages (mostly derived from monocytes). It is therefore an important helper in phagocytosis of foreign invaders, such as bacteria, as well as devouring fragments of broken erythrocytes.

The Tonsils: "Little Almonds" In the Back of Our Throat

In addition to full-blown lymphatic organs, there are smaller masses of lymphatic tissue scattered here and there around the body. Prominent among these are the *tonsils* (**TAHN**-sils). The tonsils are literally "little almonds" (*tonsils*) – oval, somewhat almond-shaped clusters of lymphatic tissue – lying in the back of the throat (Figure 17.4).

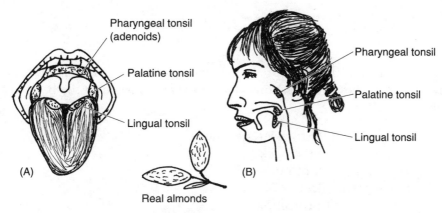

Fig. 17.4 The tonsils: "Little almonds" tucked away in our throats. (A) Frontal view. (B) Side view.

In humans, there are five tonsils. The *pharyngeal* (fah-**RIN**-jee-al) *tonsil* is the single uppermost mass, located in the portion of the *pharynx* (**FAIR**-inks) or "throat" just behind the nose. The pharyngeal tonsil is also called the *adenoids* (**AD**-uh-**noyds**), because it is rather big and "gland" (*aden*) "like" (*-oid*).

The two *palatine* (**PAL**-ah-**tyn**) *tonsils*, as their name indicates, are a pair of tonsils lying on either side of the throat, just below the *palate* (**PAL**-aht) or "roof of the mouth." Finally, there is a pair of *lingual* (**LING**-gwal) *tonsils*, attached way back at the base of the "tongue" (*lingu*).

TONSILLITIS

Since they are composed of lymphatic tissue, these five tonsils play minor roles in body defense. They contain lymphocytes and macrophages that phagocytose foreign invaders entering the nose, mouth, or throat. These invaders naturally include various airborne bacteria and viruses.

Sometimes, however, the lymphatic tissue of the tonsils becomes overwhelmed by a huge number of bacteria or viruses. In such cases, *tonsillitis* (**tahn**-sihl-**EYE**-tis) may result. Tonsillitis is "an inflammation and swelling of" (-*itis*) the tonsils. This inflammation may be accompanied by a dangerously high fever. The operation of *tonsillectomy* (**tahn**-sihl-**EK**-toh-mee) or "removal of" (-*ectomy*) the tonsils is then frequently performed.

[**Study suggestion:** Using your growing knowledge, write a single term that literally means, "inflammation of the adenoids." Why do you think that a person afflicted with this condition might have trouble breathing?]

1, Disorder

Quiz

Refer to the text in this chapter if necessary. A good score is at least 8 correct answers out of these 10 questions. The answers are listed in the back of this book.

1. Immunity literally translates to mean:
    (a) "Soreness of the back"
    (b) "A condition of not serving"
    (c) "Open to infection"
    (d) "Guards of protection"

2. The word, lymph, derives its name from what specific characteristic?
    (a) Dark red color of the fluid
    (b) Clear, watery fluid contents
    (c) Walking with a "limp" when the lymph circulation is blocked
    (d) Milky, murky sludge

3. The lymph is mainly created due to the process of:
    (a) Simple diffusion
    (b) Phagocytosis
    (c) Antigen–antibody reactions
    (d) Pressure-driven filtration

4.   The ultimate destination of all lymph:
     (a) Small masses of lymphatic tissue
     (b) Extremely large lymph veins
     (c) Blood-containing arteries
     (d) Blood-containing veins

5.   Act as scouts that first chemically detect and signal the presence of a foreign invader:
     (a) Plasma cells
     (b) Thymic lymphocytes
     (c) Antigen–antibodies
     (d) B-lymphocytes

6.   A bone marrow lymphocyte:
     (a) Readily differentiates into a plasma cell
     (b) Is a major producer of antibodies
     (c) Usually changes into a T-cell
     (d) Basically is identical to an erythrocyte

7.   Amoeba-like cells that are active in achieving immunity:
     (a) B-lymphocytes
     (b) Plasma cells
     (c) Erythrocytes
     (d) Wandering macrophages

8.   An endocrine gland that secretes thymosin:
     (a) Spleen
     (b) Pancreas
     (c) Thymus
     (d) Thyroid

9.   Both the spleen and red bone marrow are involved in the functions of:
     (a) Hematopoiesis, recycling of old erythrocytes, and phagocytosis
     (b) Antigen–antibody reactions, only
     (c) Digesting important types of foodstuffs
     (d) Body movement and support of the vertebral column

10. The type of tonsil also known as the adenoids:
     (a) Lingual
     (b) Palatine
     (c) Zygomatic
     (d) Pharyngeal

The Giraffe ORDER TABLE for Chapter 17
(Key Text Facts About Biological Order Within An Organism)

1. _____

2. _____

3. _____

4. _____

The Dead Giraffe DISORDER TABLE for Chapter 17
(Key Text Facts About Biological Disorder Within An Organism)

1. _____

18

The Respiratory System: Breath of Life

Chapter 17 talked about the lymphatic system and its circulation. But Chapter 17, which deals with the *respiratory* (**RES**-pir-ah-**tor**-ee) *system*, has much more in common with the cardiovascular or blood circulatory system.

Respiration versus Ventilation

Respiration literally means "the act of" (*-tion*) "breathing" (*spir*) "again" (*re-*). In actual usage, however, respiration is the process of gas exchange between two or more body compartments. Consider, for example, *external respiration* in mammals (Figure 18.1). External respiration is gas exchange (such as that of O_2 gas and CO_2 gas) that occurs between the blood in the pulmonary capillaries and the air within the *alveoli* (al-**VEE**-oh-lie). In mammals, each *alveolus* (al-**VEE**-oh-lus) is a "little cavity" (*alveol*) or microscopic air sac within the lung.

Internal respiration, in contrast, is gas exchange that occurs between the blood in the systemic capillaries and the fluid within the tissue cells. Internal respiration, then, is the way mammalian cells acquire oxygen and give off carbon dioxide.

During breathing, air is inhaled from the atmosphere, and it enters the millions of alveoli within the lungs. We know that breathing, itself, involves *pulmonary functions*: that is, those "pertaining to" (*-ary*) the "lungs" (*pulmon*). An appropriate term here is *ventilation* (**ven**-tih-**LAY**-shun), the "process of" (*-tion*) "fanning or blowing" (*ventil*). A ventilation system in a large building, for example, blows stale air out, and sucks fresh air in. A similar situation exists for the lungs. Therefore, *pulmonary ventilation* is the sucking of air into the lungs, followed by the blowing of air out of the lungs.

INSPIRATION VERSUS EXPIRATION

The technical terms concerning pulmonary ventilation are *inspiration* (**IN**-spir-ay-shun) on the one hand, *expiration* (**EKS**-pir-**ay**-shun) on the other hand. Inspiration is the process of sucking air into the lungs, while expiration is the process of blowing air out of the lungs.

Summarizing these relationships, we can say that:

PULMONARY VENTILATION = INSPIRATION + EXPIRATION

while

RESPIRATION = The Process of Gas Exchange between Various
               Body Compartments

1, Order

Pulmonary ventilation is the way that the body takes in fresh air (by inspiration) and gets rid of stale air (by expiration). And respiration (external and internal) is the means by which O_2 and CO_2 are exchanged between various compartments while the air is within the body.

These very basic inter-relationships are diagrammed for humans and other mammals within Figure 18.1.

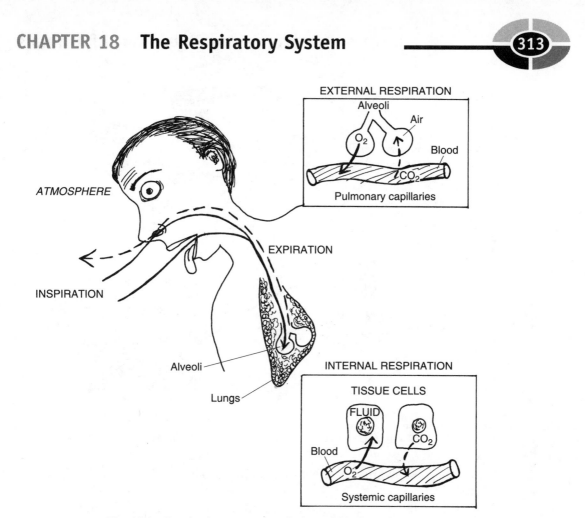

Fig. 18.1   Respiration versus ventilation within humans.

Fish Gills: Respiration and Ventilation through the Water

Land-dwelling animals – mammals, birds, reptiles, and amphibians – use air as a means for both respiration and ventilation. And associated with this characteristic, they have developed lungs, which inflate and deflate with air.

Aqueous (water-dwelling) animals, such as fish, however, ventilate water through their mouths (Figure 18.2). The water passes through the pharynx (throat), and then out across a set of curved *gill arches*. The gill arches consist of a series of thin, thread-like *gill filaments*. Thin plates on the surfaces of the gill filaments serve as the actual points of respiration (gas exchange). Water passes through the narrow gaps between the plates and moves in a single

1, Web

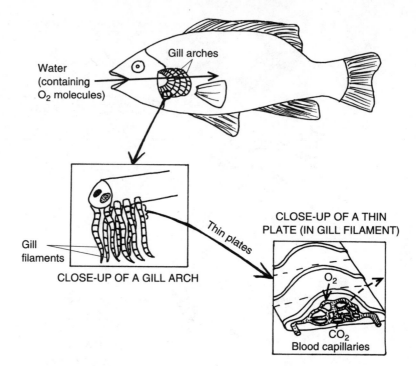

Fig. 18.2   Fish gills: Extracting oxygen from the water.

direction, out towards the edge of the _operculum_ (oh-**PER**-kyuh-lum) or gill "cover."

As the water passes, O_2 molecules diffuse out of the water, and into tiny blood capillaries. Simultaneously, CO_2 molecules diffuse out of the fish's bloodstream, and into the water. This mechanism for gas exchange is very effective. A limiting problem, however, is that the gill arches are very soft and tend to easily collapse. There is also no defense mechanism to protect gill arches in fish from dehydration. [**Study suggestion:** Ask yourself the following question: "So, why can't most fish just use their gills and extract O_2 from the air? Why can't most fish breathe outside of the water?" Use the preceding hints in your reading to help you.]

The Path of Airflow in Humans

"When humans take in air by inspiration, what is the sequence of body structures through which the air passes?" the curious-minded reader might well ask. The answer is depicted within Figure 18.3.

THE UPPER RESPIRATORY PATHWAY

Lying above both lungs is the *upper respiratory pathway*. This pathway begins with two cavities – the *nasal* (**NAY**-sal) *cavity* and the *oral* (**OR**-al) *cavity*.

The nasal cavity lies within the "nose" (*nas*), while the oral cavity is situated behind the "mouth" (*or*). These are the first two body structures that inhaled air usually enters. From both of these cavities, the air goes back into the pharynx (throat).

Situated at the lower end of the pharynx, one sees the *larynx* (**LAIR**-inks). The larynx or "voice box" is a box-shaped collection of cartilage plates held together by dense fibrous connective tissue. Like the bow of a ship, the *laryngeal* (lah-**RIN**-jee-al) *prominence* is a projection of cartilage sticking out from the front of the voice box. Stretched across the interior of the larynx are the two *vocal* (**VOH**-kal) *cords*. The vocal cords are two straps of highly elastic connective tissue, which vibrate with the passage of air through the larynx. These vibrations create the "voice" (*voc*) sounds.

The *glottis* (**GLAHT**-is) is the name of the tapered, "tongue" (*glott*)-shaped opening between the two vocal cords. Closely related to the glottis is the *epiglottis* (**EH**-pih-**glaht**-is). The epiglottis is a highly flexible flap of cartilage literally located "upon" (*epi-*) the glottis. The epiglottis thus serves as a flexible lid over the top of the larynx or voice box. When a person swallows, the food or liquid normally pushes the epiglottis shut. This usually prevents food or liquid from entering the larynx and the rest of the respiratory pathway.

Below the larynx is the *trachea* (**TRAY**-kee-ah) or main "windpipe." The trachea is stiff and noncollapsible, due to the presence of horseshoes of cartilage within its walls.

THE LOWER RESPIRATORY PATHWAY

As the trachea branches, the *lower respiratory pathway* is created. The *right* and *left primary bronchi* (**BRAHN**-kigh), the first branches, then enter the two lungs. The primary bronchi just keep branching. Eventually, a set of *bronchioles* (**BRAHN**-kee-**ohls**) or "little bronchi" emerges. Much like an inverted (upside down) tree, the *respiratory tree* thus consists of a succession of ever-smaller and more numerous branches – the larynx, trachea, bronchi, and bronchioles.

At the furthest tips of the bronchioles, hang clusters of thousands of alveoli. (Picture the many olives suspended from the branches of an inverted olive tree.) Each alveolus is essentially a collapsible, extremely thin-walled air sac. It is across the walls of the alveoli, and their close neighbors, the pulmonary capillaries, that external respiration occurs. By this means, remember, oxygen contained in the inhaled air finally enters the bloodstream.

[**Study suggestion:** Look very carefully at Figure 18.3. Which structure mentioned in this section do you think represents the so-called "Adam's apple" in males? Why?]

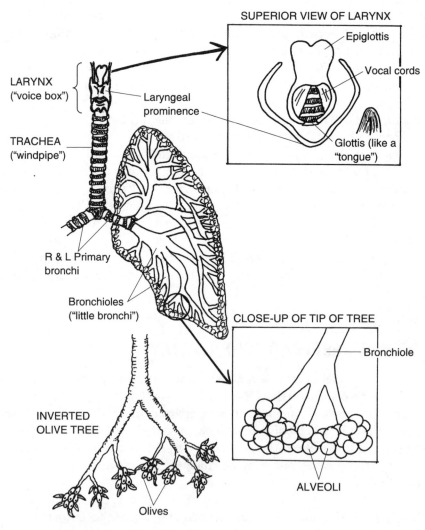

SUPERIOR VIEW OF LARYNX

Epiglottis

Vocal cords

LARYNX ("voice box")

Laryngeal prominence

Glottis (like a "tongue")

TRACHEA ("windpipe")

R & L Primary bronchi

Bronchioles ("little bronchi")

CLOSE-UP OF TIP OF TREE

Bronchiole

INVERTED OLIVE TREE

ALVEOLI

Olives

Fig. 18.3   The respiratory tree.

Frogs versus Mammals: The Mechanism of Inspiration

The previous section basically told us about the sequence of structures through which air passes when we breathe. The next obvious question we could ask is, "Okay, but how do we breathe, in the first place? How do we inhale air into our lungs?" The answer to this question involves a discussion of the mechanism of inspiration.

BULK FLOW OF AIR

Air circulates through the respiratory tree in much the same way that blood circulates through the cardiovascular system. As we learned in Chapter 17, blood flows down a blood pressure (BP) gradient, from a place where the BP is higher, towards another place where the BP is lower. In general, we can call this a *bulk flow* process. Bulk flow is the pressure-driven movement of some fluid substance (such as blood or air) from an area of greater pressure towards an area of lower pressure. In other words, bulk flow occurs down a pressure gradient.

2, Order

For ventilation in lung-breathing animals, such as frogs and mammals, then, the problem becomes one of creating an *air pressure gradient* – a difference in air pressure – between the air in the atmosphere and the air within the alveoli (tiny lung air sacs). With such a gradient, there will be a bulk flow of air from the atmosphere, and into the lung alveoli.

EQUAL GAS PRESSURES

There are two total gas pressures to consider. The *atmospheric pressure* is the pressure created by all of the gases in the atmosphere. (The atmosphere is the approximately 1 mile-thick blanket of air covering the surface of the Earth.) At sea level, the atmospheric pressure pushes with a total force of about 760 mmHg. The atmospheric pressure also pushes with this force upon the lips and nostrils. So when a person opens his mouth, the atmospheric pressure tends to push air down into his lung alveoli (see Figure 18.4, A).

Conversely, the *intra-alveolar* (**in**-trah-al-**VEE**-oh-lar) *pressure* is the total pressure exerted by all of the gas molecules within the alveoli. When a person opens his mouth, the intra-alveolar pressure tends to push air out of the alveoli, and out of the nose and mouth.

The atmospheric pressure (tending to push air into the alveoli) and the intra-alveolar pressure (tending to push air out of the alveoli) are thus two opposing pressures. Between breaths, we have an equality of these two pressures:

Atmospheric pressure = Intra-alveolar pressure

and

No air pressure gradient, therefore no bulk flow of air into or out of the lungs

POSITIVE-PRESSURE BREATHING

3, Order

In order for the frog or mammal to breathe, the equality between the atmospheric and intra-alveolar pressures must be broken. In *positive-pressure breathing*, an air pressure gradient is created by doing something "positive": that is, by increasing one of the pressures to make it higher than the other one. Consider the case of the frog and most other kinds of amphibians (Figure 18.4, B). The frog opens its nostrils, lowers the floor of its mouth, and gulps air into its oral cavity. The next step is the one that creates a positive-pressure breathing effect: the frog closes its nostrils and raises the floor of its mouth. This action creates a pushing force, thereby increasing the atmospheric pressure within the frog's mouth. Inspiration occurs because air from the mouth is then pushed down through the pharynx and trachea, and into the lung alveoli, which have a lower pressure.

NEGATIVE-PRESSURE BREATHING

2, Web

In contrast to amphibians, most reptiles, and all birds and mammals, utilize *negative-pressure breathing* for inspiration (Figure 18.4, C). The *diaphragm* (**DIE**-ah-**fram**) or "barrier" muscle forming the floor of the chest cavity contracts, and then lowers. This makes the *thoracic* (thor-**ASS**-ik) or "chest" (*thorac*) cavity larger at the bottom end, since its floor has been lowered. Since the thoracic cavity becomes larger, the lungs follow suit and enlarge along with it. And, because the lungs have enlarged, the millions of tiny alveoli become larger air sacs, as well. Finally, because the alveoli are larger, their limited number of contained gas molecules are pushing against the walls of a much larger sac. As a result, the intra-alveolar pressure does something "negative" – it falls below the atmospheric pressure, so that air is sucked into the lung alveoli by inspiration.

SUMMARY OF INSPIRATION

To summarize, positive-pressure breathing is pushing breathing, in which inspiration results from increasing atmospheric pressure above the intra-alveolar pressure; in contrast, negative-pressure breathing is suction breathing. Inspiration occurs because the intra-alveolar pressure is decreased below the atmospheric pressure, so that air is sucked into the alveoli. (Picture water being sucked down a bathtub or sink drain, due to a lower pressure inside the drain pipes.)

In humans and mammals, expiration is usually passive, involving no further spending of energy. The diaphragm relaxes, and the series of events that caused inspiration is simply reversed. Air is blown out of the mouth and nostrils. [**Study suggestion:** Examine Figure 18.4, C, and just reverse the direction of the events. You will then see for yourself the mechanism of passive expiration.]

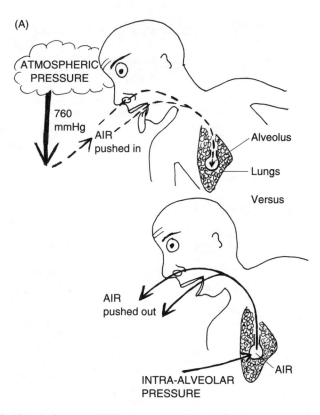

Fig. 18.4   Inspiration in frogs and little bears. (A) Atmospheric pressure versus intra-alveolar pressure. (B) Positive-pressure breathing. (C) Negative-pressure breathing.

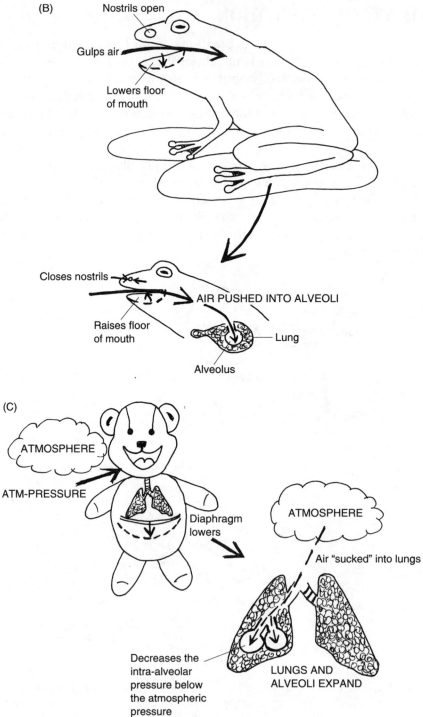

(B)

Nostrils open

Gulps air

Lowers floor of mouth

Closes nostrils

AIR PUSHED INTO ALVEOLI

Raises floor of mouth

Lung

Alveolus

(C)

ATMOSPHERE

ATM-PRESSURE

Diaphragm lowers

ATMOSPHERE

Air "sucked" into lungs

Decreases the intra-alveolar pressure below the atmospheric pressure

LUNGS AND ALVEOLI EXPAND

Major Lung Volumes and Capacities

The previous section basically told us how we breathe. This section now reveals how much we breathe. Specifically, this section deals with various *lung volumes* and *capacities*.

TIDAL VOLUME (TV)

Take a single breath while you are resting. This breath is called the *tidal volume*, abbreviated as *TV*. The tidal volume (TV) is named for its resemblance to a real tide – the moving of waves of water back-and-forth, back-and-forth, back-and-forth, upon the sand of a beach. The tidal volume is the amount of air exhaled after the person takes a normal resting inspiration. It amounts to about 500 ml (milliliters) in an average human adult.

VITAL CAPACITY (VC)

Another important measurement of pulmonary function is the *vital capacity*, abbreviated as *VC*. *Vita* (**VEE**-tah) comes from the Latin and means "life." The vital capacity (VC) therefore represents a person's capacity for life. Technically, the VC is defined as the total amount of air that a person can inhale and exhale from normal, uncollapsed lungs.

RESIDUAL VOLUME (RV)

When you or I exhale, our alveoli normally don't completely collapse. If we picture each alveolus as a balloon, then the alveolus balloon only partially deflates during expiration, simply becoming smaller, rather than totally deflating. As a result, there is usually a *residual volume*, or *RV*. The residual volume (RV) is literally the residual or "left-over" volume, still remaining within the alveoli even after expiration has occurred. This partial, residual inflation greatly reduces the amount of work required to completely re-inflate the lungs during the next cycle of inspiration.

TOTAL LUNG CAPACITY (TLC)

If we add both the vital capacity (VC) and the residual volume (RV) together, we obtain the *total lung capacity* (*TLC*):

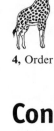

4, Order

$$\text{Total lung capacity} = \text{Vital capacity} + \text{Residual volume}$$
$$\text{(TLC)} \qquad\qquad \text{(VC)} \qquad\qquad \text{(RV)}$$

The total lung capacity therefore represents the total amount of air that the lungs can possibly hold.

Control of Respiration and Body Acid–Base Balance

The next logical question for the inquiring mind to consider is this: "Why do we breathe? What is the normal stimulus or goad that causes our diaphragm to contract, thereby resulting in inspiration?"

The answer is direct. The major stimulus for inspiration is a slight increase in the carbon dioxide (CO_2) concentration and acidity (H^+ ion) level within our bloodstream. As Figure 18.5 (A) reveals, carbon dioxide produced from cell metabolism/respiration (Chapter 4) quickly reacts with H_2O within our erythrocytes. This reaction results in H_2CO_3 as its product. H_2CO_3 is the symbol for the chemical compound *carbonic* (car-**BAH**-nik) *acid*. The carbonic acid molecules quickly breakly down into *hydrogen ions*, symbolized as H^+, and *bicarbonate* (buy-**CAR**-buh-nut) *ions*, symbolized as HCO_3^-.

Hydrogen (H^+) ions, and carbonic acid (H_2CO_3) are both classified as *body acids*, substances that either donate hydrogen ions (such as H_2CO_3) or that consist of hydrogen ions (such as a pool of H^+ ions, itself).

5, Order

During *normoventilation* (**NOR**-moh-ven-tih-**LAY**-shun) – breathing at a "normal" rate and depth – the person exhales just enough CO_2 to prevent *acidosis* (**AH**-sih-**DOH**-sis). Acidosis is an "abnormal condition of" (*-osis*) too much body "acid." Because the person exhales just enough CO_2, there isn't time for too much CO_2 to accumulate within the erythrocytes and build up too much carbonic acid, H_2CO_3, or H^+ ions. Acidosis, therefore, is presented, and a healthy state of *acid–base balance* is achieved.

HYPERVENTILATION AND ALKALOSIS

1, Disorder

"What happens to the body if you blow off too much CO_2?" the curious reader may ponder. Blowing off or exhaling too much CO_2 is what happens during *hyperventilation* (**HIGH**-per-ven-tih-**LAY**-shun). Hyperventilation is the act of breathing at an "above normal or excessive" (*hyper-*) rate and depth for current metabolic conditions. A person under great and sudden stress, for example, may fall into a state of *emotional hyperventilation*. Since they are hysterically crying and sobbing, they are breathing too hard and fast for their current metabolic condition – a state of rest rather than exercise. When this person hyperventilates (Figure 18.5, **B**), too much CO_2 is exhaled from the body. Thus, not enough CO_2 is left to react with H_2O inside the erythrocytes. And there is not enough carbonic acid (H_2CO_3) or hydrogen ion (H^+) produced.

The resulting state is *alkalosis* (**AL**-kah-**LOH**-sis). Technically speaking, alkalosis is an "abnormal condition of" (*-osis*) not enough body acid, or too much *base* or *alkali* (**AK**-kah-**lye**). In general, a base or alkali is a H^+ ion acceptor. (Bases and alkali will be discussed in more detail, along with the digestive system, in Chapter 19.) If a person hyperventilates for too long, body acid levels fall way below their normal range, and a state of alkalosis follows. The person may well get dizzy and pass out.

HYPOVENTILATION AND ACIDOSIS

"What happens if you blow off or exhale too little CO_2?" the inquiring brain may once again wonder. Blowing off or exhaling too little CO_2 is what happens during *hypoventilation* (**HIGH**-poh-ven-tih-**LAY**-shun). Hypoventilation is the act of breathing at a "below normal or deficient" (*hypo-*) rate and depth for current metabolic conditions. [**Study suggestion:** Try to name some specific situations where a person who was breathing at a rate suited for normal resting conditions, was now no longer ventilating sufficiently.]

2, Disorder

A person hypoventilating long enough may fall into a state of acidosis (Figure 18.5, **C**). There is just too much CO_2 accumulating within red blood cells, due to tissue metabolism. Not being blown off fast enough, too many CO_2 molecules react with H_2O, and too many carbonic acid molecules and hydrogen ions result. Acid–base balance is broken, and a disordered condition of acidosis reigns supreme.

Interestingly enough, acidosis (like alkalosis) may disturb brain function enough to cause the person to pass out, or even collapse into a coma!

RESPIRATION ACID–BASE SUMMARY

To summarize, we can say that:

Normoventilation = Blow off just enough CO_2 = A state of acid–base balance

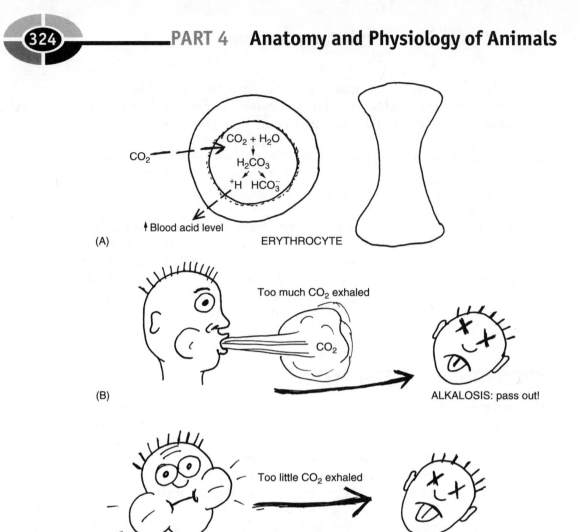

Fig. 18.5 Respiration and body acid–base balance. (A) Body acid creation in RBCs. (B) Hyperventilation ("excessive" breathing). (C) Hypoventilation ("deficient" breathing).

Hyperventilation = Blow off too much CO_2 = A state of alkalosis
Hypoventilation = Blow off too little CO_2 = A state of acidosis

When a person is stimulated to inhale by a slight rise in blood acidity, then, he or she also exhales just enough CO_2 during expiration, such that a healthy state of acid–base balance is maintained. And, as a significant bonus, enough oxygen is delivered to the brain and other vital organs during inspiration to maintain normal functioning of the entire body.

Quiz

Refer to the text in this chapter if necessary. A good score is at least 8 correct answers out of these 10 questions. The answers are listed in the back of this book.

1. Respiration actually is the process of:
    (a) Sucking air into the lungs
    (b) Converting glucose into ATP by tissue cells
    (c) Blowing air out of the lungs
    (d) Gas exchange between two or more body compartments

2. Ventilation is the process of:
    (a) Gas exchange between one body compartment and another
    (b) Removing acid from the intestine
    (c) Blowing out stale air, and sucking in fresh air
    (d) Creating new pulmonary tissue via mitosis

3. Gas exchange that occurs between the blood in the systemic capillaries and the fluid within tissue cells:
    (a) Hyperventilation
    (b) Acid–base balance
    (c) External respiration
    (d) Internal respiration

4. The actual places of gas exchange within the gills of fish:
    (a) Operculum
    (b) Lung surface
    (c) Bone surfaces within the gill arches
    (d) Thin plates on the gill filaments

5. The scientific name for the voice box:
    (a) Trachea
    (b) Epiglottis
    (c) Larynx
    (d) Pharynx

6. If the lower respiratory pathway can be compared to an inverted olive tree, then the "olives" on this tree are actually represented by the:
    (a) Primary bronchi
    (b) Bronchioles
    (c) Alveoli
    (d) Pulmonary capillaries

7. An increase in the intra-alveolar pressure above the total atmospheric pressure would most likely result in:
    (a) Sucking of air into the lungs
    (b) Severe hypoventilation
    (c) Blowing of air out of the lungs
    (d) Paralysis of the diaphragm muscle

8. The critical first event in human inspiration:
    (a) Collapse of the alveoli
    (b) Relaxation of the diaphragm muscle
    (c) Muscular constriction of the windpipe
    (d) Contraction of the diaphragm muscle

9. A person with normal lungs usually doesn't have to make a lot of effort to inspire more air, just after making an expiration. An important reason for this fact is that:
    (a) Each person usually has plenty of ATP to spare
    (b) The VC is so large that a regular expiration makes no difference in it
    (c) TLC = VC + RV
    (d) The residual volume keeps the alveoli partially inflated at all times

10. A key linkage between CO_2 and blood acidity is the:
    (a) Carrying of oxygen by hemoglobin
    (b) Release of nitrogen from tissue cells during their metabolism
    (c) Combination of carbon dioxide with H_2O in thousands of erythrocytes
    (d) Activation of acid-forming enzymes within blood leukocytes

The Giraffe ORDER TABLE for Chapter 18
   (Key Text Facts About Biological Order Within An Organism)

1. _____

2. _____

3. _____

4. _____

5. _____

The Dead Giraffe DISORDER TABLE for Chapter 18
  (Key Text Facts About Biological Disorder Within An Organism)

1. _____

2. _____

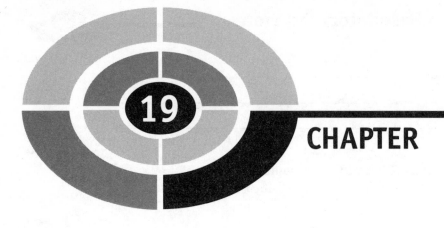

"Getting the Goodies": Nutrition and the Digestive System

Chapter 18 spoke of the "Breath of Life." By this, of course, we meant the respiratory system, and how it operates to take in life-giving oxygen and get rid of CO_2 and body acid before they accumulate to dangerous levels.

Now we will look at another life-giving process – *digestion* – and its association with *nutrition*. Nutrition is the "process of nourishing," while *nutrients* (**NEW**-tree-unts) are "nourishing (substances)." Such nutrients (nourishing substances) would include, of course, glucose and other simple sugars, lipids, and proteins consumed for energy within our diet.

The Digestive Tube and Its Basic Processes

Digestion is literally the "process of separating or dividing (something) apart." Formally speaking, digestion is the chemical and physical breakdown of food, thereby releasing its contained nutrients. But digestion and its breakdown is not an isolated process. It occurs along with many other related processes, and in the same anatomic location – within the *digestive tube* (Figure 19.1). The digestive tube is a tube that extends from the mouth (oral cavity) to the *anus* (**AY**-nus), the small, muscular "ring" through which one defecates.

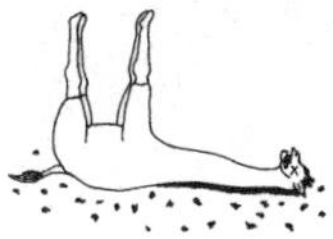

1, Disorder

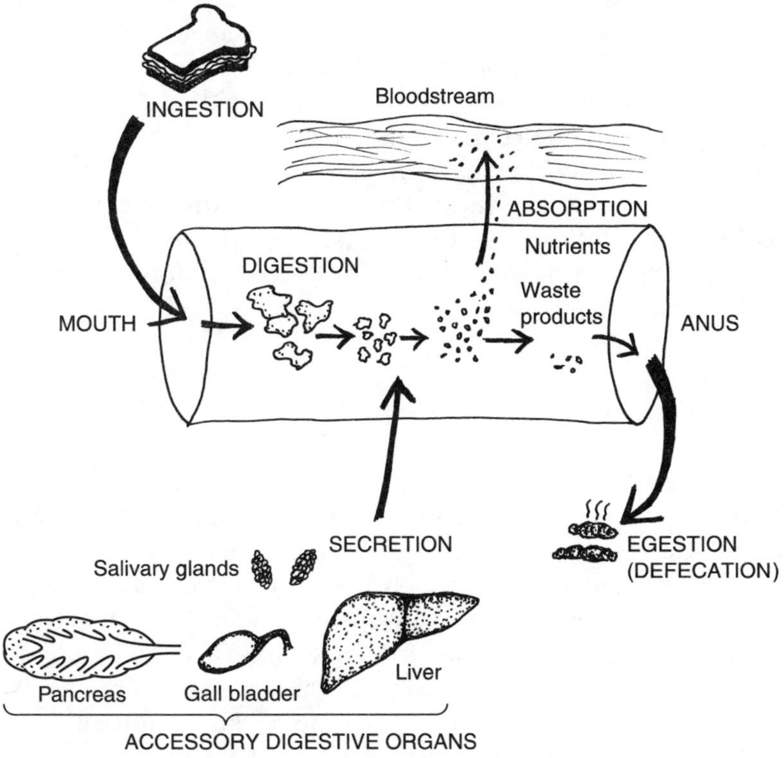

Fig. 19.1   The digestive tube and its general functions.

INGESTION AND EGESTION

Of the major processes associated with the digestive tube, *ingestion* (**in-JES**-chun) is the first, while *egestion* (**ee-JES**-chun) is the last. Ingestion is the "process of carrying (food) into" (*in-*) the digestive tube, while egestion is the "process of carrying (feces) out" (*e-*).

DIGESTION AND ABSORPTION

Absorption is the "process of swallowing (something) up." In reality, absorption is a different process from the actual one of swallowing. Swallowing is the movement of food from the oral cavity, down into the pharynx (throat), and then into the *esophagus* (eh-**SAHF**-uh-**gus**) or "gullet," the muscular tube leading into the stomach.

Absorption is the movement of digested nutrients from the interior of the digestive tube, into the bloodstream. But we must remember that, quite often, the food we ingest consists of large chunks of food, far too big to be absorbed. (Picture a ham sandwich, with protein molecules in the ham and starch or carbohydrate molecules in the bread.)

Just because a particular foodstuff happens to be ingested through the mouth, does not necessarily mean that it can be digested (broken down) into smaller nutrients and absorbed. This digestive process is somewhat dependent upon the particular kind of stomach the food is entering, and the various *digestive enzymes* it may contain.

SECRETION AND DEFECATION

Digestion is often helped by the results of *secretion* – the release of various products from *accessory digestive organs* that are added to the digestive tube contents. The accessory digestive organs are organs attached to the sides of the digestive tube, but through which no food or feces actually passes. In humans, the accessory digestive organs include the *salivary* (**SAH**-lih-**vair**-ee) *glands*, pancreas, *liver*, and *gall bladder*. These organs all add small quantities of various secretions to the digestive tube, which help in its process of breaking down big chunks of food into smaller molecules of nutrients.

Egestion is also called *defecation* (**deh**-feh-**KAY**-shun), the excretion of unusable waste products within the feces or excrement.

1, Order

SUMMARY

The five basic processes involving the digestive tube are thus ingestion, digestion, secretion, absorption, and egestion (defecation).

The Digestive Tube (Alimentary Canal) in Mammals

The digestive tube is alternately called the *alimentary* (**al**-uh-**MEN**-tur-ee) *tract*. The reason, of course, is that the digestive tube has many of its major functions "related to" (*-ary*) "nourishing" (*aliment*) the rest of the body. Let us now, therefore, examine the parts of the human digestive tube in sequence, along with their related functions. (Start looking at Figure 19.2). And in so doing, we shall also be learning about the digestive tube in other mammals.

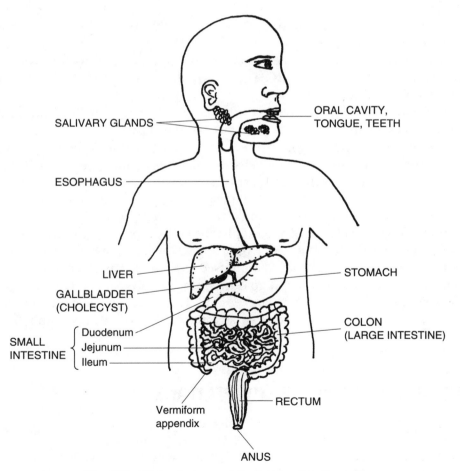

Fig. 19.2   Major features of the human digestive tube.

2, Disorder

THE ORAL CAVITY, PHARYNX, AND ESOPHAGUS

Food ingested by a human (or other mammal) immediately enters the mouth (oral cavity) where *physical digestion* of all three major types of foodstuffs (carbohydrates, lipids, and proteins) begins. Physical digestion is just the mechanical breaking apart of ingested food, using the teeth, lips, and gums.

In humans, *chemical digestion* of carbohydrates also begins within the oral cavity. This is due to the presence of "spit" or *saliva* (sah-**LIE**-vah), secreted into the mouth by the salivary glands. The saliva contains various digestive enzymes, such as *salivary amylase* (**AM**-ih-**lace**), or "starch" (*amyl*) "split-ter." Salivary amylase begins the chemical digestion (breakdown) of *complex carbohydrates*, such as starch, into double-sugars. [**Study suggestion:** Eat a plain saltine cracker. At first, it is quite dull, reflecting its complex starch content. But as you chew it and mix it with your saliva, notice that it begins to taste sweet. To what specific chemical should you give credit for this dramatic change?]

By the time a person is done chewing food and mixing it with saliva, the general result is a food *bolus* (**BOH**-lus). A food bolus is a soft "ball" (*bol*) of partially digested food.

When done chewing, the person uses the tongue and flips the food bolus into the back of the oropharynx (portion of the throat behind the mouth). The bolus pushes the epiglottis shut, then slides down into the esophagus. The upper portion of the esophagus is lined by voluntary striated (cross-striped) muscle. Hence, the first part of swallowing is voluntary. ("So, why did I just gulp down that piece of delicious apple?" you might well ask yourself. "Because I darn well wanted to, that's why!")

However, the lower 2/3 of the esophagus is lined by smooth, involuntary muscle. This means that the latter part of swallowing is not under our conscious control. Therefore, once a swallowed food bolus has entered the lower esophagus, you just have to let it go down into your stomach! ("Oh, oh!" you might suddenly question yourself. "Didn't I just see half a worm in that chunk of apple?" Too late! You've already swallowed it! You can't back out, now!)

THE STOMACH AND ITS ADAPTATIONS

In humans, the stomach is a capital J-shaped pouch that acts as a temporary storage place for ingested food. The *body* is the large central area of the stomach, while the *pylorus* (pie-**LOR**-us) or "gatekeeper" is a small room at the far end of the stomach.

The epithelial cells lining the "stomach" (*gastr*) secrete the *gastric* (**GAS**-trick) *juice*. The gastric juice is especially rich in *hydrochloric* (**HIGH**-droh-**klor**-ik) *acid*, abbreviated as *HCl*, and *pepsin* (**PEP**-sin). Being an extremely strong acid, HCl breaks down rapidly and donates many H^+ ions, making it highly reactive and corrosive. Thus, hydrochloric acid begins the chemical digestion of lipids and proteins, as well as continuing the digestion of carbohydrates. Pepsin is an enzyme that helps break down proteins, as well.

3, Disorder

Due to the action of the gastric juice, the food bolus from the esophagus is now changed into *chyme* (**KIGHM**). The chyme is a thick, soupy mass of partially digested material. Since it is almost a liquid, chyme is like a "juice" (*chym*) that leaves the stomach through a muscular ring called the *pyloric* (pie-**LOR**-ik) *sphincter*.

"If the stomach is full of so much HCl, then why doesn't it digest itself?" Part of the answer is that the stomach secretes a highly alkaline (basic) layer of mucus, a protective "slime" (*muc*) that coats the lining and neutralizes acid that contacts it.

There is one type of foodstuff that HCl and pepsin hardly touch, however. That foodstuff is *cellulose* (**SELL**-yuh-**lohs**). Cellulose is a "carbohydrate" (-*ose*) composed of the walls of many "little cells" (*cellul*) found in plants. Wood, cotton, and grass, for instance, are largely made up of cellulose. "Why don't we just send the kids out in the backyard to graze on our grass?" a sarcastic father might jokingly ask. "After all, then we wouldn't have to mow it, and they'd get big as cows!"

A comparison to the anatomic features of the cow stomach may help explain why this dad is really off the mark! As Figure 19.3 reveals, the cow stomach consists of four different chambers. These chambers are called the *rumen* (**ROO**-men), *reticulum*, *omasum* (oh-**MAY**-sum), and *abomasum* (**ab**-uh-**MAY**-sum). Because the first chamber, the rumen, is so important, it serves as the foundation for the name of the entire group of herbivores (plant-eating mammals) to which cows, deer, goats, and sheep belong. This group is called the *ruminants* (**REW**-muh-nunts) – animals that are "chewers of cud."

1, Web

Cud represents a mouthful of food that has been swallowed and then brought back up into the mouth from the rumen (first stomach chamber) of a ruminant. The cud is then given a slow, thorough, second chewing before it is swallowed, again. But simple re-chewing isn't good enough, by itself. This is because cows, like humans, don't produce enzymes capable of chemically digesting plant cellulose (as found in grass). To get around this problem, there are special bacteria and protists (Chapter 7) living in the first two stomach chambers, the rumen and reticulum. Now, the microorganisms within these chambers do have enzymes that digest cellulose in

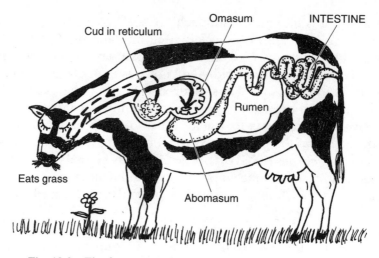

Fig. 19.3   The four stomach chambers of the cud-chewing cow.

grass and break it down into simple sugars that the cow's stomach can further handle, then absorb. The repeated chewing of cud containing millions of these micro-organisms simply crunches up the cellulose plant walls, thereby allowing the enzymes to digest them even more efficiently.

THE WONDROUS SMALL INTESTINE

Within humans, the *small intestine* is a small-diameter, extensively folded tube that averages about 20 feet (6 meters) in length. We have called this the wondrous small intestine, because of all the wonderful and amazing feats of digestion and absorption that occur within its walls.

The first segment of the small intestine is called the *duodenum* (dew-**AH**-den-um), which comes from Medieval Latin for "presence of 12." Thus, the name, duodenum, reflects the fact that ancient anatomists measured its length as equal to about 12 fingers, placed side-by-side. A close look at Figure 19.4 also shows how critical the duodenum is as a common meeting point for chyme and various digestive secretions.

The duodenum receives chyme from the pylorus of the stomach, and secretions from the liver, gall bladder, and pancreas, as well. The liver is a large, brown, many-lobed organ that produces and secretes *bile*, as well as many other useful substances. Bile is a brownish-green detergent substance that *emulsifies* (ih-**MUL**-sih-feyes) fat within the small intestine. *Emulsification* (ih-**mul**-suh-fuh-**KAY**-shun) is literally the "process of" (*-tion*) "milking out" (*emulsif*) one non-mixable fluid substance from another one. Consider, in this case, the large globules of partially digested fat that do not mix very well with the soupy chyme entering the duodenum from the stomach. Bile from the liver acts to emulsify the large fat globules, breaking them apart (or, in a sense, "milking them out" of the rest of the chyme). As a result, a separate foam of tiny fat droplets is created within the small intestine. [**Study suggestion:** Pour some liquid detergent onto a bunch of greasy plates, and then observe what happens. In what way does this liquid detergent act somewhat like bile?]

4, Disorder

Bile is secreted continuously, day and night, into the right and left *hepatic* (heh-**PAT**-ik) or "liver" *ducts*. These ducts carry the bile into the *cystic* (**SIS**-tic) *duct*. Cyst means "bladder" or "sac," while *chole* (**KOH**-lee) is Latin for "bile or gall." Hence, the compound word, *cholecyst* (**KOH**-luh-sist), translates into English as "gall bladder" or "bile sac."

The cholecyst (gall bladder) is a muscular-walled sac that receives the bile from the liver and stores it temporarily. When the duodenum becomes swollen with fatty chyme, a hormone is released that stimulates the walls of the gall bladder to contract. A load of bile is squirted out of the cholecyst, much like a slug of brownish-green pea soup or gravy being squeezed out of a rubber balloon.

The bile squirts into the cystic duct, and then into the *common bile duct*, which carries it the rest of the way down into the duodenum. Here, then, the emulsification of fat takes place.

As shown in Figure 19.4, the *pancreatic* (pan-kree-**AT**-ik) *duct* extends from the pancreas and merges with the base of the common bile duct. The pancreatic duct is the main passageway for the *pancreatic juice*. Surrounding both of them at their point of union is the *hepatopancreatic* (heh-**PAT**-oh-pan-kree-**AT**-ik) *sphincter* (**SFINK**-ter). The hepatopancreatic sphincter is a ring of smooth muscle that regulates the emptying of both the common bile duct (the hepatic or "liver" portion) and the pancreatic duct into the small intestine. When this sphincter (muscular ring) relaxes, bile and pancreatic juice flow through the *duodenal* (dew-**AH**-deh-nal) *papilla* (pah-**PIL**-lah). The duodenal papilla is a "little nipple or pimple" (*papill*) -like projection with a hole in its center. Bile from the liver and pancreatic juice from the exocrine gland portion of the pancreas (Chapter 15) drip into the duodenum through the hole in the duodenal papilla.

The pancreatic juice contains *sodium bicarbonate* (buy-**CAR**-buh-**nayt**), symbolized chemically as $NaHCO_3$, as well as a variety of digestive enzymes. These enzymes include amylases (starch-splitters), *lipases* (**LIE**-pay-sez) or "fat-splitters," and *proteases* (**PROH**-tee-**ay**-sez) or "protein-splitters." The lipases, for example, complete the chemical digestion of fat or lipids, after they have been emulsified by bile into a fatty foam. The resulting products,

5, Disorder

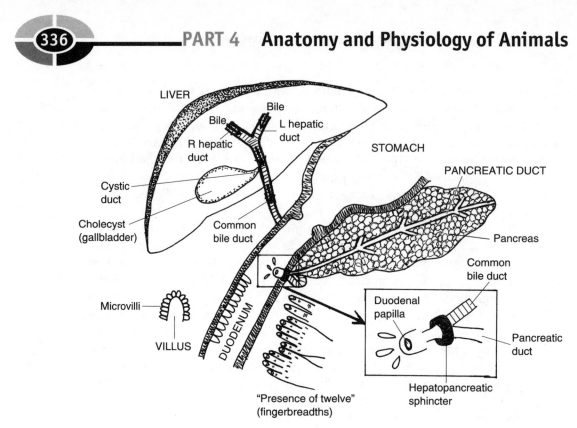

Fig. 19.4 The duodenum and its friendly neighbors.

such as *fatty acids* and the substance, *glycerol* (**GLIH**-sir-**ahl**), are then absorbed across the walls of the small intestine, and into the bloodstream. Similarly, the proteases continue the chemical breakdown of proteins into amino acids, which are also absorbed into the bloodstream. And the amylases in the small intestine generally finish the chemical breakdown of carbohydrates into simple sugars such as glucose, which are then absorbed.

Reflecting its critical role in the absorption of nutrients, there are several important modifications to the *mucosa* (mew-**KOH**-sah), the "mucous" (*mucos*) membrane lining the small intestine. The mucosa of the duodenum, jejunum, and ileum is thrown into thousands of *villi* (**VIL**-ee). Review of Figure 19.4 shows that each single *villus* (**VIL**-us) is literally named for its resemblance to a little bump or curved "tuft of hair" (*vill*). The surface of each villus, in turn, is covered with dozens of *microvilli* (**MY**-kroh-**vil**-ee). The microvilli are little bumps upon each villus, like "tiny tufts of hair." The numerous villi and microvilli, by throwing the mucosa up into hundreds of tiny bumps, vastly increases the amount of surface area available for absorption. As a result, the absorption of nutrients into the bloodstream from the small intestine is extremely efficient.

2, Order

Revisiting the liver once more, we can say that it has numerous other critical body functions, in addition to the production of bile. First among these is *detoxification* (dee-**TAHKS**-ih-fih-**KAY**-shun) – "the process of taking poison out of" the bloodstream. The liver cells help *detoxify* (dee-**TAHKS**-ih-feye) drinking alcohol, for instance, so that it is broken down into sugar and water without *toxifying* ("poisoning") the brain! [**Study suggestion:** When we say that a person is drunk, what does that imply, with regards to the associated function of the liver?]

Summarizing all of the above, we can say that the duodenum, as the first segment of the small intestine, receives chyme from the stomach, bile from the liver and cholecyst, and pancreatic juice from the pancreas. As a result, the chemical digestion of all three basic types of foodstuffs – carbohydrates, lipids, and proteins – is essentially completed within the small intestine. Lying downstream from the liver and pancreas entry point — the duodenal papilla — are the *jejunum* (jeh-**JOO**-num) and the *ileum* (**IL**-ee-um).

The jejunum and ileum basically complete the processes of chemical digestion and absorption of nutrients that began in the duodenum.

We thus have the following simple summary equation:

$$\text{SMALL INTESTINE} = \text{Duodenum} + \text{Jejunum} + \text{Ileum}$$

3, Order

THE COLON: OUR LARGE INTESTINE

The last major section of the digestive tube is the *colon* (**KOH**-lun) or "large intestine." The colon (large intestine) is a wide-diameter, folded tube, about 6 feet (2 meters) in length in an average-sized human adult (see Figure 19.5). The colon begins with the *cecum* (**SEE**-kum), a "blind" (*cec*) or dead-ended pouch that has the *vermiform* (**VER**-mih-form) *appendix* or "worm-like attachment" hooked to its base. The vermiform appendix is basically a solid attachment of modified lymphatic tissue (Chapter 17) that plays a minor role in the body's immune or self-defense system.

Liquid chyme from the ileum of the small intestine pushes through the *ileocecal* (**il**-ee-oh-**SEE**-kul) *sphincter*. Once within the cecum, the chyme begins to undergo an extensive drying out process, wherein large amounts of water and salt (H_2O) are absorbed. In addition, there are beneficial bacteria in the colon that produce a variety of B-vitamins, as well as sulfur-containing amino acids, which are also absorbed.

Due to this drying out process, chyme is solidified into *feces* (**FEE**-seez) within the colon. Besides H_2O, feces also contain a significant percentage of *fecal* (**FEE**-kal) *bacteria* and *dietary fiber* (actually undigested cellulose material).

Looking at Figure 19.5, you might well ask yourself, "How does chyme/ feces in the cecum, which is way down at the bottom of the large intestine, move up to the other parts of the colon?" Good question! The answer is, by *mass peristalsis* (pear-ih-**STAHL**-sis). Peristalsis literally means "a constriction" (*-stalsis*) "around" (*peri-*). Formally defined, then, peristalsis is a constriction (narrowing) around a particular point of some tube, due to the contraction of a circular ring of smooth muscle around the tube. Mass peristalsis is thus the simultaneous constriction or narrowing of a large number of directly neighboring points along the large intestine wall, such that the heavy mass of the feces is pushed on to the next section of the colon. [**Study suggestion:** Either in your imagination, or for real, get a large tube of toothpaste. Turn it upside down, and remove the cap. Now, make a ring around the upper part of the tube using both your hands. Count one . . . two . . . three . . . . Constrict your fingers hard around the tube! What do you observe? This constriction at a single point is a crude model of peristalsis. Think. How could you use the same basic elements to model mass peristalsis in the colon? If you phoned up several of your friends, what could you do?]

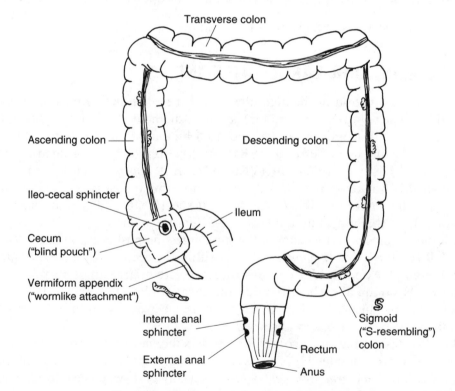

Fig. 19.5 The human colon (large intestine).

After the cecum, next in sequence are the *ascending colon* (which goes upward), *transverse colon* (which runs sideways), and the *descending colon* (which goes downward). *Sigmoid* (**SIG**-moyd) means "S-resembling." Hence, the *sigmoid colon* is the S-resembling portion of the large intestine coming right after the descending colon.

The sigmoid colon snakes down into the *rectum* (**WRECK**-tum), a muscular-walled, "straight" (*rect*) tube that empties feces into the anus. There are two sphincters within the rectum. The higher one, called the *internal anal sphincter*, is not under our conscious control. The internal anal sphincter tends to open whenever mass peristalsis has moved some feces down into the upper portion of the rectum.

Fortunately, for us, there is also a ring of voluntary striated (cross-striped) muscle, positioned in the lower portion of the rectum. This muscular ring is called the *external anal sphincter*. Its contraction and relaxation is very much under our conscious control (at least, ever since we were first "potty-trained")! Therefore, we can usually choose the time and place where we will consciously relax this lower sphincter and carry out defecation (egestion).

Let us end by summarizing in words the parts of the large intestine:

COLON = Cecum + Ascending colon + Transverse colon
+ Descending colon + Sigmoid colon + Rectum

4, Order

Too Many, or Too Few Calories?

Way back in Chapter 4 we talked about the Chemical Balance of Life, and its relationship to cellular respiration or metabolism. Recall that in the catabolism (breakdown) of chemical foodstuffs (such as glucose molecules), free or kinetic energy is released and tied to the production of more cell ATP. Obviously, these topics are closely related to our current discussion, since digestion is, in fact, defined as the chemical and physical breakdown of food. Similarly, nutrition is the process of nourishing or feeding the body cells.

THE CONCEPT OF CALORIES

The nutrients that humans eat, say, in a candy bar, are all associated with a certain number of *calories* (**KAL**-or-**ees**). A calorie is technically a unit for describing the amount of "heat" (*calor*) that is released during the breakdown of food. In nutrition, we generally use the "large calorie" or *kilocalorie* (**KILL**-uh-**kal**-ur-ee) – the amount of heat energy released measured in "thousands" (*kilo-*) of calories. A candy bar, for instance, may contain 670 kilocalories of heat energy, when it is consumed.

THE CONCEPT OF CALORIC BALANCE

5, Order

An important principle of Biological Order in humans and other animals is the notion of *caloric* (kuh-**LOR**-ik) *balance*. Caloric balance is a condition where the number of kilocalories consumed in food per day exactly equals the number of kilocalories burned in exercise per day. Summarizing, we have:

CALORIC BALANCE :   Number of kilocalories consumed/day
= Number of kilocalories burned/day

Caloric balance is desirable, therefore, whenever we wish to maintain our current body weight. It logically follows that whenever one wishes to either gain or lose weight, then some degree of Biological Disorder must be inserted into the scheme, such that the caloric balance is disrupted.

When the number of kilocalories consumed per day significantly exceeds the number burned during exercise, there is a net excess of kilocalories. Some of these excess kilocalories are stored as extra glycogen deposits in liver and muscle cells, and some of it is stored as extra lipid within our adipose tissue cells. The net result, of course, is that we (gulp!) gain weight and often get fatter!

Conversely, to lose weight, the number of kilocalories burned during daily exercise must exceed the number consumed in foodstuffs (no matter what the current fad diet books tell you!).

Quiz

Refer to the text in this chapter if necessary. A good score is at least 8 correct answers out of these 10 questions. The answers are listed in the back of this book.

1. For humans, eating grass would not be a good way of obtaining nutrients, because:
    (a) Grass contains no nutrient molecules
    (b) Blades of grass cannot be digested by any known type of animal

   (c) Human stomachs and small intestines lack the key enzymes
       necessary to break down cellulose
   (d) Grass can't even be chewed by human beings!

2.   Ingestion is the exact opposite of what process?
   (a) Egestion
   (b) Digestion
   (c) Absorption
   (d) Mastication

3.   The movement of nutrients from the interior of the digestive tube into
   the blood:
   (a) Absorption
   (b) Secretion
   (c) Defecation
   (d) Ingestion

4.   The chemical digestion of carbohydrates begins in the mouth, owing to
   the presence of:
   (a) Hydrochloric acid
   (b) $NaHCO_3$
   (c) NaCl
   (d) Salivary amylase

5.   A thick, soupy mass of partially digested material found in the stomach
   and small intestine:
   (a) Bolus
   (b) Cud
   (c) Chyme
   (d) Feces

6.   The first segment of the small intestine:
   (a) Jejunum
   (b) Vermiform appendix
   (c) Duodenum
   (d) Cecum

7.   Technical term for the gallbladder:
   (a) Gallicule
   (b) Cholecyst
   (c) Cystic duct
   (d) Hepatic accessory organ

8. Regulates the emptying of both the common bile duct and the
    pancreatic duct into the duodenum:
    (a) Pyloric sphincter
    (b) Microvillus
    (c) Hepatopancreatic sphincter
    (d) Ileo-cecal sphincter

9. The large intestine begins with the dead-ended _____:
    (a) Sigmoid colon
    (b) Rectum
    (c) Cecum
    (d) Transverse colon

10. A critical structure that allows defecation to be voluntary:
    (a) Vermiform appendix
    (b) Internal anal sphincter
    (c) Descending colon
    (d) External anal sphincter

The Giraffe ORDER TABLE for Chapter 19
  (Key Text Facts About Biological Order Within An Organism)

1. _____

2. _____

3. _____

4. _____

5. _____

The Dead Giraffe DISORDER TABLE for Chapter 19
   (Key Text Facts About Biological Disorder Within An Organism)

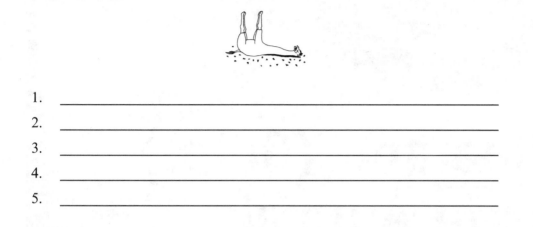

1. _____
2. _____
3. _____
4. _____
5. _____

The Spider Web ORDER TABLE for Chapter 19
   (Key Text Facts About Biological Order Beyond the Individual Organism)

1. _____

Urine and (Gulp!) Sex in Animals

Chapter 19 told us all about digestion and nutrition, two essential functions for the survival of the individual organism. Now, in Chapter 20, we consider both the *urinary* (**YOUR**-ih-**nair**-ee) and *reproductive* (**ree**-proh-**DUCK**-tiv) *systems*. The urinary system literally "pertains to" (-*ary*) "urine" production, storage, and excretion from the individual body. The reproductive system in the male and female, on the other hand, is literally about "producing" a new organism, "again" (*re-*). For most animals, this implies using the *genital* (**JEN**-ih-tal) *organs* to "beget or produce" (*genit*) sexually.

The Genitourinary (Urogenital) System Concept

In humans and other mammals, it is appropriate to speak not just of the urinary and reproductive systems alone but of a combined *genitourinary*

(**JEN**-ih-toh-**ur**-ih-**nair**-ee) or *urogenital* (**UR**-oh-**jen**-ih-tal) *system*. This is because many of the structures of the urinary and reproductive (genital) organs are shared in common. Consider, for example, the *penis* (**PEA**-nis) in males. The penis is a spongy "tail" (*pen*)-like structure that serves both to carry urine out of the body, as well as deliver spermatozoa (sperm cells) to an ovum (mature egg cell) for reproduction.

1, Order

Major Urinary Structures in Animals

In all vertebrates, the major organs of urine excretion are the *kidneys*. In humans, a pair of bean-shaped kidneys are located along either side of the vertebral column, deep within the back.

1, Web

KIDNEY ANATOMY

Figure 20.1 provides an overview of *renal* (**REE**-nal) or "pertaining to" (*-al*) "kidney" (*ren*) anatomy. The kidney is encased within the *renal capsule*, a thin membrane of fibrous connective tissue. The kidney, itself, is subdivided into three major areas or zones. The outermost zone is called the *renal cortex*. Much as the cerebral cortex forms a thin "bark" over the surface of the cerebrum (Chapter 14), the renal cortex does the same for the kidney. The "middle" (*medull*) area is the *renal medulla* (meh-**DEW**-lah). And the deepest zone is the *renal pelvis* (**PEL**-vis). The renal pelvis is a broad, bowl-shaped sac that receives the urine as it flows from the renal cortex and medulla. And carrying the collected urine of the renal pelvis is the *ureter* (**YOUR**-eh-**ter**).

2, Order

Within the renal cortex are millions of *nephrons* (**NEF**-rahns). The nephrons are the major microscopic functional units of the kidney. It is the nephrons that are actually responsible for formation of urine from the blood. Each nephron begins with a *glomerulus* (gluh-**MAHR**-yew-**lus**). The glomerulus is a tiny, red-colored collection of *renal capillaries*. This structure gets its name from its resemblance to a little red "ball of yarn" (*glomerul*). The blood pressure pushing against the walls of the capillaries in each glomerulus, causes a filtration of fluid out of the glomerulus, and into the adjoining group of *urinary tubules* (**TWO**-byools) – "tiny urine tubes."

The urinary tubules from each group of neighboring nephrons eventually empty into a common passageway called a *collecting duct*. A number of collecting ducts pass down together through the renal medulla. They create the *renal pyramids*, which are pointed at their bottom tips like the rather

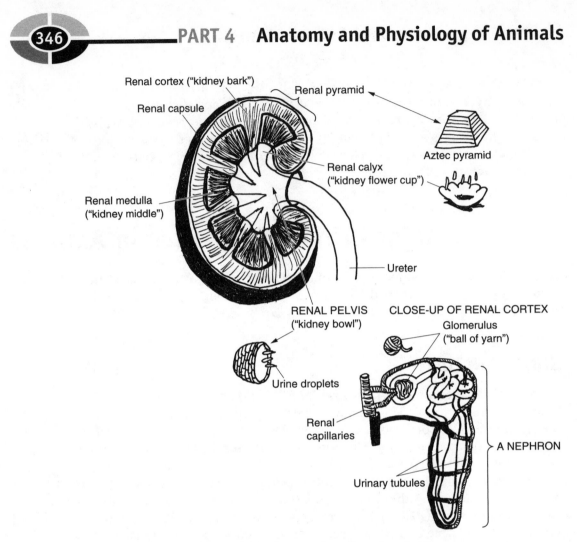

Fig. 20.1   An overview of renal anatomy.

blunt pyramids constructed by the Aztecs or Inca Indians. The tip of each renal pyramid drips urine into a *renal calyx* (**KAY**-licks), or "kidney flower cup." And the urine from each calyx eventually flows into the body of the renal pelvis, before it leaves the kidney via the ureter.

THE URINARY PATHWAY

Figure 20.2 shows the rest of the urinary pathway, lying beyond the kidney. The right and left ureters both dump urine into the *urocyst* (**YUR**-oh-**sist**) or *urinary bladder* (*cyst*). The urocyst (urinary bladder) is a hollow, muscular-walled pouch that temporarily stores the urine before it is excreted.

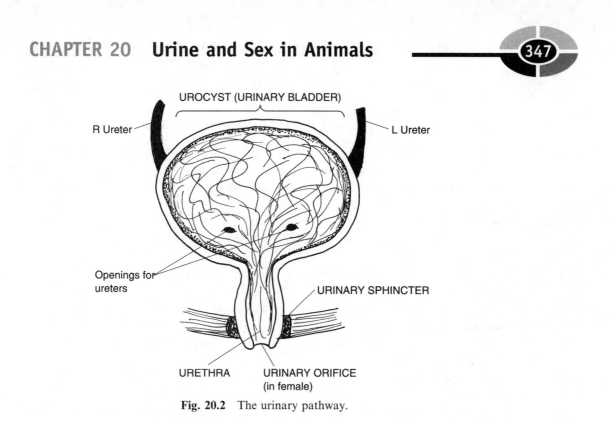

Fig. 20.2   The urinary pathway.

The urocyst empties into the *urethra* (you-**REETH**-rah), the tube that helps a person literally "make water" (*urethr*) – that is, *urinate* (**YUR**-ih-**nayt**). Surrounding the upper neck of the urethra is the *urinary sphincter*. Much like the external anal sphincter in the digestive pathway (Chapter 19), the urinary sphincter is a ring of voluntary striated muscle. This means, of course, that the contraction and relaxation of this sphincter is under our voluntary control. Thus, after we have been adequately "potty-trained" during early childhood, we can voluntarily relax the urinary sphincter whenever the place and time are right for urination!

Finally, urine exits out of the body through the *urinary orifice* (**OR**-ih-**fis**), a tiny, "mouth" (*or*)-like opening.

The Process of ("Making Water") or Excreting Urine

We have, from time to time, been using some very basic descriptive equations in this book. Let us do so, again. We can state the *urinary excretion equation*, for example:

$$E = F - R + S$$

(This equation is put into visual form within Figure 20.3.)

In this equation, **E** stands for "excretion," **F** for amount filtered, **R** for the amount of *tubular reabsorption*, and **S** for *tubular secretion*. The urinary *filtrate* (**FIL**-trayt), or filtration product, comes from the pushing force of the blood pressure against the walls of the renal capillaries in the *glomeruli* (glah-**MEHR**-you-**lie**). This quantity of filtrate is huge, averaging about 180 liters of fluid per day, in an average adult! [**Study suggestion:** Assume that an adult has a blood volume of about 6 liters. Then, on average, how many times is this person's entire blood volume filtered out of his glomeruli, each day?]

While it may seem wasteful, the huge volume of urinary filtrate (**F**) acts as the starting point for the urine. Because there is so much of this filtrate, the body can adjust many factors to influence how much urine is actually excreted, under particular current conditions.

After urinary filtration, one of the chief processes is tubular reabsorption (**R**). Reabsorption is the movement of material out of the filtrate, across the walls of the urinary tubules, and back into the bloodstream. Consider, for example, the tubular reabsorption of glucose. Under normal conditions, almost 100% of the glucose that is filtered into the urinary tubules is eventually reabsorbed back into the bloodstream. As a result, the urine excreted from the body is nearly free of glucose. Several hormones also control the amount of sodium (Na^+ ions) and water (H_2O) molecules that are reabsorbed back into the bloodstream. Because the amount of sodium (salt) and water reabsorbed can vary greatly, the kidneys play a critical role in regulating the *salt–water balance* of the human bloodstream.

Under typical conditions, about 99%, or 179 liters, of the urinary filtrate (mostly water) is reabsorbed. [**Study suggestion:** If a person becomes extremely dehydrated, as after excessive sweating, then what do you predict will happen to the amount of H_2O reabsorbed? Will the percent (%) reabsorbed increase above typical conditions, or decrease below it? Why?]

Another process, tubular secretion (**S**), involves the active (ATP-requiring) addition of small quantities of particular chemicals from the bloodstream, into the urinary tubules. Molecules of *penicillin* (pen-ih-**SILL**-in) and many other *antibiotics* (**an**-tih-buy-**AH**-ticks), for instance, are just too large to be filtered across the walls of the glomeruli. Hence, the epithelial cells lining the blood vessels actively pump the penicillin into the urinary tubules. Therefore, penicillin is excreted (**E**) out of the body, via the urine. Although only a few milliliters (ml) of fluid are generally secreted each day, they still have an important influence.

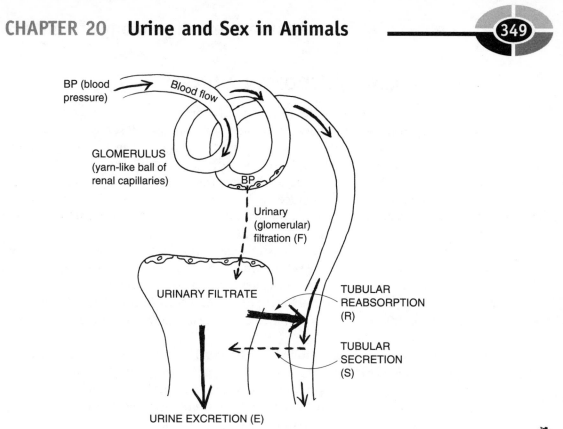

Fig. 20.3 Processes involved in excretion of urine.

Summarizing our previous urinary excretion equation and plugging in some numbers, we get:

$$\underset{\text{1 L/day of urine excreted}}{\mathbf{E}} \underset{=\ 180\ \text{L/day}}{=\ \mathbf{F}} \underset{-\ 179\ \text{L/day}}{-\ \mathbf{R}} \underset{+\ (\text{a few ml/day})}{+\ \mathbf{S}}$$

3, Order

Reproduction in Animals: Sexual or Asexual?

Earlier chapters have talked about sexual reproduction occurring in animal organisms and in non-animal organisms such as in the World of Plants (Chapter 9). Even in reproducing pine trees, there are sperm cells from the male (pollen) cones that fertilize egg cells (ova) from the female (seed) cones, thereby resulting in a zygote. The lowly fungi (Chapter 8) were likewise shown to engage in sexual reproduction – the fusion of a male and female gamete (sex cell) to create a zygote. And this zygote in turn develops into an embryo, within many different types of organisms.

2, Web

ASEXUAL REPRODUCTION IN ANIMALS

But we also need to remember that asexual reproduction is likewise common! In this reproductive strategy, there are no gametes or sex cells involved. Asexual reproduction, you may remember, is important in the spreading of the black bread mold, Rhizopus. (Review Figure 8.4, if desired.)

3, Web

Certain animals (living organisms that are neither plants nor fungi) also use asexual means to reproduce themselves. Not surprisingly, the animals engaging in asexual reproduction are generally among the most primitive. Consider, for example, the invertebrates (Chapter 10). Certain types of star-fish reproduce asexually by means of *fragmentation* (breaking of their body into fragments), followed by *regeneration* (the re-growing of lost body parts). When such a starfish has one of its arms fragmented (broken off), an extensive series of *mitoses* (my-**TOH**-seez) or cell divisions occur within the broken arm fragment. One such broken arm can thus asexually reproduce (via mitosis) a whole new starfish!

Sea anemones (also mentioned in Chapter 10) can asexually reproduce by means of *fission* (**FIH**-shun) – "the process of" (*-ion*) "splitting or cleaving" (*fiss*). In this case, one approximately round sea anemone stretches into two identical individuals, creating a type of dumbbell-shaped pattern. The two duplicate individual sea anemones then split or cleave apart.

Finally, some invertebrates such as tunicates asexually reproduce by *budding*. A new individual simply grows as a small bud off the body of a parent. When the growing bud becomes large enough, it separates from the parent body.

Basically, all of the above examples of asexual reproduction have two main features in common:

1. The major mechanism for reproduction is simply an extensive series of mitoses (cell divisions involving no reduction in the number of chromosomes present).
2. A group of *clones* (**KLOHNS**) – identical copies of a single parent organism – is created. Like a branch or "twig" (*clon*) growing from the trunk of the same tree, a clone has an exact copy of the genes and chromosomes of its parent.

EXTERNAL FERTILIZATION: SEXUAL REPRODUCTION IN LOWER VERTEBRATES

In animals engaging in sexual reproduction, the mature ovum of the female is fertilized by a sperm cell from the male. In certain types of lower (more primitive) vertebrates, such as fish and amphibians, there is *external fertilization* of the ovum "outside of" the body of the female.

Consider two mating frogs. The larger female frog is mounted by a smaller male, who clasps her body with his forelimbs. This helps stimulate the female to release a mass of over 100 eggs into the pond water! Nearly simultaneously, the male frog releases a jet of sperm cells, which fertilize many of the ova externally, right in the surrounding water. After a time of development, the fertilized ova develop into swimming tadpoles.

4, Web

INTERNAL FERTILIZATION: SEXUAL REPRODUCTION IN HIGHER VERTEBRATES

When one examines the mating of higher vertebrates such as mammals, sexual reproduction generally occurs via *internal reproduction*. In this type of reproduction, the ovum is fertilized by mature sperm cells that have traveled "within" (*intern*) the female body.

5, Web

Reproductive System of the Male Mammal

In order to fully understand internal fertilization, we must first study the reproductive system of the male and female mammal (in particular, of human beings).

Figure 20.4 reveals the basic anatomy of the reproductive system in the human male. A *scrotum* (**SKROH**-tum) or "leathery bag of skin" (*scrot*) suspends the two *testes* (**TES**-teez) outside of the abdominal cavity. Each *testis* (**TES**-tis) is a rather oval, whitish, "eggshell" (*test*)-like structure that contains the *seminiferous* (sem-ih-**NIF**-er-us) *tubules*. The seminiferous tubules are a collection of tiny, highly coiled tubes that carry out the process of *spermatogenesis* (sper-mat-uh-**JEN**-eh-sis) – the "production of" (-*genesis*) "sperm" (*spermat*) cells.

From the time of puberty (age 12–13 years) onward, mature sperm cells are continually produced by a *germinal* (**JER**-muh-nal) *epithelium*, which is located in the thick walls of the seminiferous tubules. This germinal epithelium undergoes a constant process of "sprouting" (*germin*) new sperm cells by mitosis, followed by meiosis.

You may recall (Chapter 9) that meiosis is literally "a conditioning of lessening." This implies that division of a parent cell by meiosis lessens or reduces the number of chromosomes in each resulting daughter cell by one-

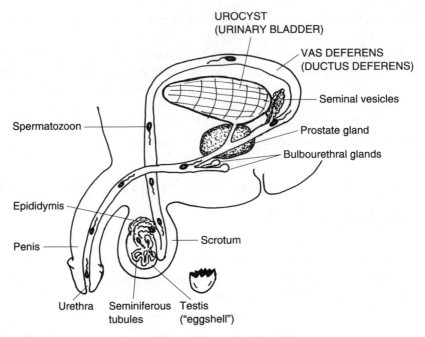

Fig. 20.4   The male reproductive pathway.

half. In the human male, each primitive sperm cell contains 46 chromosomes within its nucleus. But after meiosis, the developing sperm cell has this reduced by half, to a total of just 23 chromosomes. Eventually, a mature *spermatozoon* (sper-**mat**-uh-**ZOH**-un) or "seed" (*sperm*) "animal" (-*zoon*) with only 23 chromosomes results.

Thousands of mature *spermatozoa* (**sper**-mah-tah-**ZOH**-ah) or "seed animals" leave the germinal epithelium of the seminiferous tubules, and are temporarily stored within the *epididymis* (**eh**-pih-**DID**-ih-mus). The epididymis is a curved, comma-shaped pouch that literally lies "upon" (*epi-*) each testis or "eggshell" (*didym*). The thousands of spermatozoa are ejected from the epididymis during each *ejaculation* (ih-**JACK**-yuh-**lay**-shun). Ejaculation is literally the "throwing out" (*ejacul*) of *semen* (**SEE**-mun) and spermatozoa from the urinary orifice at the tip of the penis.

ACCESSORY REPRODUCTIVE ORGANS IN THE MALE

The semen is a thick, milky, sugar-rich, very basic fluid that suspends the spermatozoa and gives them nutrition. When the male has an *orgasm* (**OR**-gaz-um), or is literally "swollen and excited," he ejaculates spermatozoa suspended in a fluid of semen. The stored spermatozoa are actively sucked out of the epididymis by strong peristalsis (ring-like muscular contractions) of the walls of the *vas* (vas) *deferens* (**DEF**-er-enz).

The vas deferens, or "carrying away" (*deferens*) "vessel" (*vas*), is not actually a blood vessel at all. More appropriately, it is alternately called the *ductus* (**DUCK**-tus) *deferens* or "carrying away duct." During male orgasm, the walls of the vas deferens (ductus deferens) powerfully and rhythmically constrict or narrow. This negative pressure (suction) event draws the stored spermatozoa out of the epididymis, carrying them over the top of the urinary bladder and out into the urethra.

Semen is added to the spermatozoa from a number of *accessory male reproductive organs*. These accessory male reproductive organs include the two *seminal* (**SEM**-ih-nal) *vesicles*, the two *bulbourethral* (**BUL**-boh-you-**REE**-thral) *glands*, and the single *prostate* (**PRAH**-state) *gland*.

Reproductive System of the Female Mammal

In order to achieve internal fertilization, the penis of the male must be inserted into the *vagina* (vah-**JEYE**-nah), a tapered "sheath" (*vagin*) that also serves as the birth canal. The vagina leads into the uterus or womb through a tiny hole in its neck-like *cervix* (**SIR**-viks) (see Figure 20.5).

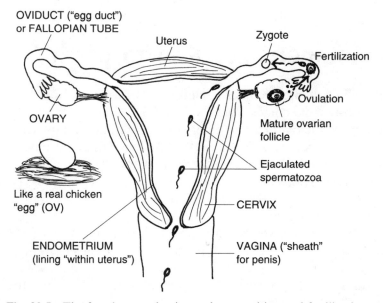

Fig. 20.5 The female reproductive pathway and internal fertilization.

Attached to the top of the uterus are the right and left *oviducts* (**OH**-vih-ducts) or "egg ducts." The oviducts are alternately called the *Fallopian* (fah-**LOH**-pea-un) *tubes* in honor of their discoverer, the Italian anatomist, Gabriello Fallopio (fah-**LOH**-pea-oh). The oviducts (fallopian tubes) are a pair of slender egg ducts that carry ova towards the uterus.

The ova are released from the *ovaries* (**OH**-var-**eez**). There is both a right ovary and a left ovary, each named for their oval, whitish appearance, much like a chicken "egg" (*ovari*).

INTERNAL FERTILIZATION AND THE FEMALE REPRODUCTIVE PATHWAY

The key to internal fertilization, of course, is both ejaculation by the male and ovulation by the female. (You may want to review appropriate parts of Chapter 15, the glands, at this point.) Previously, we learned that increased secretion of luteinizing hormone was the usual trigger for ovulation. Basically, the LH (luteinizing hormone) dissolves and weakens the outer surface of the ovary, allowing a mature ovarian follicle to rupture and release an ovum.

The ovum is usually swept up into a nearby oviduct (Fallopian tube). Fertilization usually occurs in the first (outer) 1/3 of the oviduct. Fusion of sperm and ovum creates a zygote. As the zygote moves through the oviduct and towards the uterus, it undergoes a series of *mitoses* (my-**TOH**-seez).

Embryo Development Leading to Birth

4, Order

The single-celled zygote becomes a solid mass of cells called a *morula* (mor-**OO**-luh), whose name comes from the Latin for "little mulberry." The morula passes out of the oviduct and enters the *body* (main hollow cavity) of the uterus. Here it becomes a *blastula* (**BLAS**-chew-lah), alternately called a *blastocyst* (**BLAS**-toh-**sist**). The blastula (blastocyst) is a "little sprouter" (*blastul*) or "hollow sprouting bladder" (*blastocyst*) that embeds itself within the *endometrium* (en-doh-**ME**-tree-um), the "inner" (*endo-*) epithelial lining of the "uterus" (*metr*).

After the blastula (blastocyst) implants itself into the *uterine* (**YOU**-ter-in) wall, it starts a process of cellular differentiation (the cells becoming different from each other and specialized). This leads to a *gastrula* (**GAS**-true-lah) or hollow "little stomach" (*gastrul*). As Figure 20.6 clearly displays, the gastrula gives rise to the three *primary germ layers* of the human embryo.

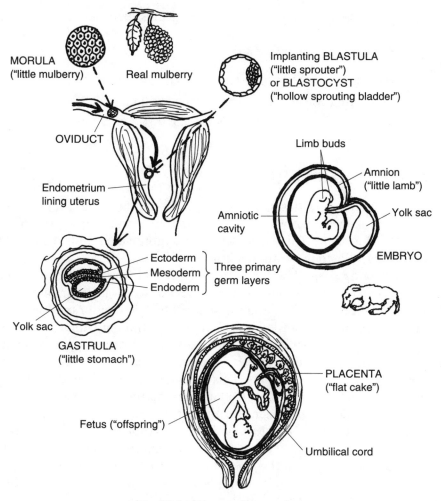

MORULA
("little mulberry")

Real mulberry

Implanting BLASTULA
("little sprouter")
or BLASTOCYST
("hollow sprouting bladder")

OVIDUCT

Endometrium
lining uterus

Limb buds

Amnion
("little lamb")

Amniotic
cavity

Yolk sac

EMBRYO

Ectoderm
Mesoderm } Three primary
Endoderm } germ layers

Yolk sac

GASTRULA
("little stomach")

PLACENTA
("flat cake")

Fetus ("offspring")

Umbilical cord

Fig. 20.6   From embryo to fetus.

The gastrula does, indeed, have some of the same characteristics as a miniature version of the real stomach! Besides having a hollow interior, the gastrula (like the real stomach), has multiple layers of cells within its walls. These cell layers are technically called *derms* – various layers of "skin."

The *endoderm* (**EN**-doh-**derm**), or "inner skin," for instance, eventually creates much of the inner lining of the main body cavities in the mature adult. The *mesoderm* (**ME**-so-**derm**) – "middle skin" – differentiates to become bone and muscle tissue. And the *ectoderm* (**EK**-toh-**derm**) specializes to ultimately become much of the skin and nervous system. It is the skin and nervous system (Chapter 14), after all, that lies at or near the body surface,

5, Order

reacts to environmental stimuli, and communicates extensively about such stimuli.

To summarize the above information, we have:

THREE PRIMARY GERM LAYERS OF THE EMBRYO =
Endoderm ("Inner skin") + Mesoderm ("Middle skin") +
Ectoderm ("Outer skin")

FROM EMBRYO TO FETUS

The *embryo*, in general, represents a "sweller." In humans, the embryo represents all of the stages of body development that occur during the first 3 months of development. This starts with the zygote stage and includes the morula, blastula (blastocyst), and gastrula.

The embryo body keeps "swelling" by extensive mitoses, adding thousands and thousands of new tissue cells. The embryo body becomes surrounded by an *amnion* (**AM**-knee-un). The amnion is a protective membrane forming a sac around the embryo of "little lambs" (*amnions*), human beings, and other mammals. An amnion also protects the embryos of other types of higher vertebrates such as birds and reptiles. The amnion also encloses a quantity of *amniotic* (**am**-knee-**AH**-tik) *fluid*. This fluid creates a watery cushion and shock absorber for the embryo, as well as keeping the body wet and moist.

A *yolk sac* provides nourishment for the early embryo. *Limb buds* develop and mark the locations of the future body limbs. An *umbilical* (um-**BILL**-ih-kal) *cord* stretches from the "central pit" (*umbilic*) area present in the mid-section of the embryo, out to the *placenta* (plah-**SEN**-tah). The placenta is a "flat cake" (*placent*)-like organ that supplies nourishment to the later embryo and fetus as the yolk sac progressively shrinks in size.

The fetus is the stage of the "offspring" from the third month after fertilization, up to the time of birth. The "newborn" child is literally a *neonate* (**KNEE**-oh-**nayt**).

Sexual Dysfunctions and Reproductive Failure

1, B-Web

A *sexual dysfunction* (dis-**FUNK**-shun) is a type of "bad, painful, or difficult" (*dys-*) failure of reproduction. There are many specific types of sexual dysfunction – some involving the male reproductive pathway, some involving the female reproductive pathway – so that successful production of a zygote is prevented. The normal patterns of Biological Order leading to fertilization are either interrupted or blocked.

MALE SEXUAL DYSFUNCTION

One important type of sexual dysfunction in males is *infertility* (**in**-fer-**TILL**-ih-tee) – the inability to produce living spermatozoa that can fertilize an ovum. One fairly common cause of infertility is *cryptorchidism* (**krip-TOR**-kid-**ih**-zum). Cryptorchidism is a "condition of" (*-ism*) "hidden" (*crypt*) "testes" (*orchid*). Cryptorchidism occurs when one or both of the testes fail to descend out of the abdominal cavity, and into the scrotum. If such *undescended testes* remain within the abdominal cavity for too long a period after birth, the high internal body temperature may kill the germinal epithelium in the wall of the seminiferous tubules. Thus, no fertile spermatozoa are produced.

1, Disorder

FEMALE SEXUAL DYSFUNCTION

Females, too, can suffer from infertility, among many other types of possible sexual dysfunctions. Female infertility is often due to either hypersecretion (excessive secretion) or hyposecretion (deficient secretion) of particular hormones. Consider, for example, the important hormone, thyroxine. Recollect (Chapter 15) that thyroxine helps regulate the basal metabolic rate (BMR), the rate at which body cells operate and burn calories under basal (resting) conditions.

2, Disorder

In *hypothyroidism* (**HIGH**-poh-**THIGH**-royd-izm), there is a "deficient or below normal" (*hypo-*) activity of the thyroid gland. As a result, hyposecretion of thyroxine slows down the metabolism of the female ovaries. When the ovaries are seriously underactive, ovulation may be postponed or even blocked. Female sterility is one unfortunate result.

Quiz

Refer to the text in this chapter if necessary. A good score is at least 8 correct answers out of these 10 questions. The answers are listed in the back of this book.

1. The reproductive organs are essentially the same thing as the _____ organs:
    (a) Urinary

    (b) Genital

    (c) Digestive

    (d) Musculoskeletal

2.   The outermost zone of the kidney:

    (a) Renal pelvis

    (b) Adrenal medulla

    (c) Renal cortex

    (d) Glomerulus

3.   The major microscopic functional unit of the kidney:

    (a) Nephron

    (b) Renal capsule

    (c) Renal pyramid

    (d) Collecting duct

4.   A ring of voluntary striated muscle that allows us to consciously control urination:

    (a) Cardiac sphincter

    (b) Cecum

    (c) Glomerular ring structure

    (d) Urinary sphincter

5.   A tiny yarn-like ball of renal capillaries:

    (a) Urinary tubule

    (b) Renal pyramid

    (c) Glomerulus

    (d) Nephron

6.   Fragmentation and regeneration is an example of:

    (a) Urine formation in certain invertebrate animals

    (b) Sexual reproduction in particular starfish and tunicates

    (c) Embryo development in most types of fish

    (d) A type of asexual reproduction

7.   A female frog mounted by a male discharges her eggs into the water. This process is part of:

    (a) Internal fertilization

    (b) Asexual reproduction

    (c) External fertilization

    (d) In vitro (test-tube) fertilization

8. An eggshell-resembling structure that contains the seminiferous tubules:
    (a) Germinal epithelium
    (b) Scrotum
    (c) Epididymis
    (d) Testis

9. The main mechanism of ejaculation in males:
    (a) Gravity-influenced fall of spermatozoa from the penile orifice
    (b) A simple diffusion of spermatozoa from a higher to lower concentration
    (c) Powerful contracting force of ductus deferens peristalsis pulls sperm out of storage
    (d) The epididymis explodes due to over-filling with sperm cells

10. A pair of slender egg ducts that carry ovulated ova towards the uterus:
    (a) Fallopian tubes
    (b) Peyer's patches
    (c) Ureters
    (d) Seminal vesicles

The Giraffe ORDER TABLE for Chapter 20
   (Key Text Facts About Biological Order Within An Organism)

1. _____

2. _____

3. _____

4. _____

5. _____

The Dead Giraffe DISORDER TABLE for Chapter 20
  (Key Text Facts About Biological Disorder Within An Organism)

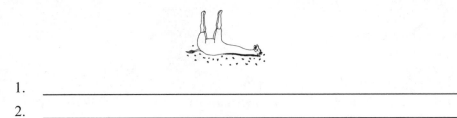

1. _____

2. _____

The Spider Web ORDER TABLE for Chapter 20
  (Key Text Facts About Biological Order Beyond the Individual Organism)

1. _____

2. _____

3. _____

4. _____

5. _____

The Broken Spider Web DISORDER TABLE for Chapter 20
  (Key Text Facts About Biological Disorder Beyond the Individual Organism)

1. _____

Test: Part 4

DO NOT REFER TO THE TEXT WHEN TAKING THIS TEST. A good score is at least 18 (out of 25 questions) correct. Answers are in the back of the book. It's best to have a friend check your score the first time, so you won't memorize the answers if you want to take the test again.

1. The skin and skeleton can be legitimately studied together because:
   (a) They both arise from two neighboring germ layers in the embryo
   (b) Each is composed of hard bony tissue
   (c) Neither is the exact opposite of the other
   (d) Organ systems never die
   (e) Oftentimes both types of systems are missing in late embryos

2. Substance mainly responsible for waterproofing human skin:
   (a) Keratin
   (b) Melanin
   (c) Glycogen
   (d) Albumen
   (e) Chlorophyll

3. Possess an endoskeleton comprised of bone and joint connective tissue:
   (a) Most arthropods
   (b) Some crabs

(c) A few types of fish, only
(d) Most spiders
(e) Humans and most other vertebrates

4. Maintenance of blood calcium homeostasis is important because:
(a) Ca^{++} is essential for all enzyme function
(b) Bone matrix cannot store many calcium ions
(c) Contractions of all body muscles depend upon an adequate supply of blood Ca^{++}
(d) Homeostasis can never be violated
(e) People's bones are easily fractured when there is too much body calcium

5. Within a bone–muscle lever system, the _____ usually serves as the fulcrum:
(a) Tendon
(b) Joint
(c) Synovial membrane
(d) Ligament
(e) Bursa

6. Perimysium is located where within a skeletal muscle organ?
(a) Around each fascicle of muscle fibers
(b) Lying upon the entire muscle organ
(c) Within each bundle or fascicle of fibers
(d) Between fascicles, but within the muscle organ
(e) Neither within nor outside of muscle fiber fascicles

7. According to the sliding filament theory, muscle fibers contract because:
(a) Thin actin myofilaments slide inward over tilted myosin cross-bridges
(b) Thick myosin myofilaments become longer, causing the actin to shorten
(c) Myofibrils pile upon one another, somewhat like an accordion
(d) Thick myosin myofilaments slide outward over vertical actin cross-bridges
(e) Muscle proteins are progressively destroyed and digested

8. Spinal nerves and individual peripheral nerves are considered parts of the:
(a) CNS
(b) Spinal cord
(c) Cerebrum

(d) PNS

(e) Neuromuscular junction

9. Sensory receptors are specialized nerve endings that are sensitive to a particular kind of:
   (a) Thought impulse
   (b) Stimulus
   (c) ACh molecule
   (d) Motor impulse
   (e) Higher motor neuron

10. Exocrine glands:
    (a) Are without ducts
    (b) Secrete major hormone products
    (c) Depend upon ducts to carry their secretions
    (d) Release messengers into the bloodstream
    (e) Include the thyroid and anterior pituitary

11. Releasing hormones are:
    (a) Secreted into the bloodstream by special neurons in the hypothalamus
    (b) Sent into a system of ducts
    (c) Produced by epithelial cells within the posterior pituitary gland
    (d) Modified by the adrenal cortex before they are actually secreted
    (e) Named for their direct influence upon the thyroid gland

12. Follicle-stimulating hormone (FSH) primarily acts to:
    (a) Release insulin into the bloodstream
    (b) Enhance the development of ovarian follicles and their secretion of estrogen and progesterone
    (c) Rupture the mature ovarian follicles, resulting in ovulation
    (d) Decrease the rate of secretion of growth hormone (GH)
    (e) Reduce the symptoms of tissue inflammation

13. Helps the adrenal medulla carry out the "Fight-or-Flight" stress response:
    (a) Corpus luteum
    (b) Secondary sex characteristic
    (c) Sympathetic portion of the nervous system
    (d) The structure of the progesterone molecule
    (e) A calm, serene, mental attitude

14. An alternate name for the cardiovascular system:
    (a) Respiratory network
    (b) Pituitary stalk

    (c) Circulatory system
    (d) Endocrine system
    (e) Integumentary system

15. The tiniest blood vessels:
    (a) Capillaries
    (b) Veins
    (c) Arterioles
    (d) Arteries
    (e) Venules

16. Have a two-chambered heart with only a single circulation:
    (a) Fish
    (b) Amphibians
    (c) Reptiles
    (d) Humans
    (e) All other mammals besides humans

17. The contraction and emptying phase of each heart chamber:
    (a) Systole
    (b) "Dupp"
    (c) Diastole
    (d) Cardiac cycle
    (e) "Lubb"

18. A below normal or "deficient" blood pressure:
    (a) Normotension
    (b) Arteriosclerotic heart disease
    (c) Hypotension
    (d) Atherosclerosis
    (e) Hypertension

19. Closely associated with immunity:
    (a) Human exoskeleton
    (b) Digestive tract
    (c) Lymphatic and reticuloendothelial system
    (d) Nodal tissue
    (e) Myocardium

20. External respiration occurs between the blood in the _____ and the air in the _____:
    (a) Systemic arterioles; pulmonary capillaries

(b) Pulmonary capillaries; alveoli
(c) Pulmonary arteries; alveoli
(d) Bronchi; trachea
(e) Systemic capillaries; bronchioles

21. Fish respiration differs from that in land-dwelling animals in that they:
    (a) Extract much more oxygen from the surrounding air
    (b) Have a better defense mechanism to protect wet breathing surfaces from dehydration
    (c) Utilize H_2O as a source for both obtaining O_2 and excreting CO_2
    (d) Take advantage of the lungs as a mechanism for ventilation
    (e) Possess five lungs, instead of just two

22. A flexible flap of cartilage forming the lid over the voice box:
    (a) Epiglottis
    (b) Primary bronchus
    (c) Glottis
    (d) Thyroid cartilage
    (e) Laryngeal prominence

23. Digestion:
    (a) The movement of food from the pharynx into the esophagus
    (b) A type of peristalsis found in the human stomach
    (c) Essentially the same thing as caloric balance
    (d) The chemical and physical breakdown of food
    (e) Passage of nutrients across the wall of the digestive tube, into the blood

24. Consists of the duodenum plus the jejunum plus the ileum:
    (a) Small intestine
    (b) Colon
    (c) Cecum
    (d) Vermiform appendix
    (e) Rectum

25. In humans, the stage of development from the third month after fertilization:
    (a) Blastula
    (b) Embryo
    (c) Yolk sac
    (d) Fetus
    (e) Zygote

Final Exam

DO NOT REFER TO THE TEXT WHEN TAKING THIS EXAM. A good score is at least 75 correct. Answers are in the back of the book. It's best to have a friend check your score the first time, so you won't memorize the answers if you want to take the test again.

1. Modern biology is literally the
   (a) "Removal of organisms"
   (b) "Study of life"
   (c) "Examination of objects under the microscope"
   (d) "Study of Nature"
   (e) "Love of knowledge"

2. Biological Order essentially represents
   (a) Very organized patterns seen in living things
   (b) Too much chaos in the affairs of Nature
   (c) The exact opposite of Body Organization
   (d) Contradictory things to different human observers
   (e) A deficiency of regular arrangements between body parts

3. The head, neck, and black spots of a giraffe
   (a) Microbial physiology
   (b) Simple structures

    (c) Animal functions
    (d) Animal anatomy
    (e) Plant regularity

4. The "process of cutting" the body "up or apart"
    (a) Biology
    (b) Geology
    (c) Paleontology
    (d) Bacteriology
    (e) Anatomy

5. You might consider body function as being what corresponding part of a sentence?
    (a) Verb
    (b) Adverb
    (c) Adjective
    (d) Noun
    (e) Object

6. _____ only occurs within living organisms
    (a) Gross anatomy
    (b) Body structure
    (c) Physiology
    (d) Microanatomy
    (e) Occupation of space

7. Rising of oral body temperature to a high of about 99.6 degrees F, and a falling down to a low of about 97.6 degrees F
    (a) Normal range
    (b) Changing of anatomy
    (c) Mean (average) value of oral body temperature
    (d) Absolute constancy of body temperature
    (e) Complete chaos of body temperature

8. A characteristic found only in living things
    (a) Definite patterns of structural arrangement
    (b) Metabolism and excretion
    (c) Dynamic changes in function over time
    (d) Generation of heat associated with movement
    (e) Response to changes in the environment

9. Rising of oral body temperature far beyond 99.6 degrees F, or falling of it far below 97.6 degrees F
    (a) Biological Order

(b) Homeostasis
(c) Denoted by a living giraffe with black spots
(d) Biological Disorder
(e) Stable pattern of the living internal environment

10. The study of the relationships among different organisms, and their interactions with the external environment
(a) Anthropology
(b) High school physics classes
(c) Physiology
(d) Anatomy
(e) Ecology

11. The icon of a dead spider and its web
(a) Homeostasis
(b) Biological Disorder within an organism
(c) Biological Disorder extending beyond the individual organism
(d) Maintenance of ecological systematic
(e) A high degree of Environmental Order

12. The world's first great biologist and Father of Natural History
(a) James Dean
(b) Frances Crick
(c) Plato
(d) Aristotle
(e) Little Lord Fauntleroy

13. Natural History can best be considered as:
(a) A concentration upon the "household affairs" of an organism
(b) Primarily a study of the Fossil Record
(c) The comprehensive study of all living creatures
(d) The exact opposite of Modern Biology
(e) Genetics, only

14. The _____ method starts with a hypothesis (educated guess or hunch)
(a) Natural History
(b) Experimental
(c) Geological
(d) Taxidermy
(e) Classification

15. Literally translates from Latin to mean a "control of sameness"
(a) Homeostasis
(b) Normal range

(c) All aspects of human physiology

(d) Metabolism

(e) Hypothesis

16. The Father of Experimental Physiology and originator of the concept of homeostasis
    (a) Aristotle
    (b) Gregor Mendel
    (c) Claude Duvall
    (d) Charles Darwin
    (e) Claude Bernard

17. A stacked Pyramid of Life was used in this book to symbolize the
    (a) Interactions between body structure and body function
    (b) Various levels of biological organization
    (c) States of Biological Disorder, primarily
    (d) Existence of anatomy & physiology at just a single level of complexity
    (e) Collapse of most ecosystems, given enough time

18. The simplest and smallest level of biological organization
    (a) Community
    (b) Atoms
    (c) Cells
    (d) Tissues
    (e) Subatomic particles

19. The lowest living level of biological organization
    (a) Population
    (b) Organelle
    (c) Cell
    (d) Tissue
    (e) Organism

20. A group of individuals of the same species that live together in the same place
    (a) Ecosystem
    (b) Community
    (c) Population
    (d) Organism
    (e) Tissue

21. Homeostasis is restricted to what portion of the Great Pyramid of Life?
    (a) Area beyond the community but below the ecosystem

(b) Organism and below
(c) Cell and organelle
(d) Tissues and organs
(e) Atom and below

22. A population balance maintained over time
(a) Relative constancy of an organism
(b) Homeostasis
(c) Normal range
(d) Ecological relationships
(e) Regular physiology

23. Mammals are literally
(a) Animals with milk-giving "breasts"
(b) Creatures who "walk freely"
(c) Animals with "backbones"
(d) Organisms having "internal order"
(e) Species that "reproduce" themselves using sex

24. Cosmic Order
(a) The increasing tilt and speed at which solar systems keep moving away from one another
(b) The ongoing Death of Stars
(c) Creation of Earth's solar system
(d) Early absence of life on our world
(e) Constant disturbances that may kill off an organism

25. Sometime between 3.5 billion and 4 billion years ago
(a) Dawn of Life
(b) The original "Big Bang" occurred
(c) Appearance of the Milky Way galaxy
(d) Creation of our planet
(e) First appearance of H_2O in Earth's atmosphere

26. Threw doubt upon the Spontaneous Generation Theory
(a) Charles Montgomery
(b) Francesco Reid
(c) Karl Marx
(d) Louis Pasteur
(e) Groucho Marx

27. "Life only occurs because it has been produced by other living organisms"
(a) Recessive gene hypothesis

(b) Theory of Biogenesis
(c) Theory of Evolution
(d) Law of Attraction between unlike charges
(e) Notion of a primordial soup

28. Among the oldest-known fossils
    (a) Filament-shaped ancient bacteria preserved within stromatolite rocks
    (b) Dead bee bodies encased in amber
    (c) Imprints of ferns in Western Sandstone deposits
    (d) Calcified remains of gigantic fish
    (e) Mummified skeletons of early human ape-men

29. A process where living organisms utilize the energy in sunlight to make sugar
    (a) Glycolysis
    (b) Aerobic metabolism
    (c) Photosynthesis
    (d) Chlorophyll molecules
    (e) Oxygen consumption

30. Types of cells that probably came into existence before cells with "kernels"
    (a) Vertebrate
    (b) Embryonic
    (c) Zygote
    (d) Prokaryotes
    (e) Aerobes

31. Organelle carrying out most aerobic metabolism within cells
    (a) Nucleus
    (b) Ribosome
    (c) Mitochondrion
    (d) Plasma membrane
    (e) Centriole

32. "But with the coming of multicellular creatures, _____ of body structure and function arrived"
    (a) Generality
    (b) Merging
    (c) Specialization
    (d) Destruction
    (e) Denucleation

33. Type of cell containing a nucleus surrounded by its own membrane
    (a) Stromatolite
    (b) Filament
    (c) Prokaryote
    (d) Dry
    (e) Eukaryote

34. The _____ were among the first pioneering multicellular eukaryotes
    (a) Ferns
    (b) Fungi
    (c) Algae
    (d) Sea gulls
    (e) Squids

35. Fungus
    (a) Greenish plant engaging in photosynthesis
    (b) Aerobic organism using glycolysis to obtain energy
    (c) Fertilized ovum of a mammal
    (d) Plant-like organism that feeds on living or dead organic matter
    (e) Greenish-blue anaerobes whose cells are filled with chlorophyll

36. Any "living, breathing" multicellular organism that is not a plant or fungus
    (a) Microbe
    (b) Animal
    (c) Reptile
    (d) Fish
    (e) Duck

37. Delicate jellyfish gracefully floating in the Pre-Cambrian seas
    (a) Pathogenic plants
    (b) Chordates
    (c) Vertebrates
    (d) Fungi
    (e) Invertebrates

38. Time of 500–200 million years ago, with first vertebrates, land plants, and insects
    (a) Cenozoic Era
    (b) Paleozoic Era
    (c) Mesozoic Era
    (d) Jurassic Park Period
    (e) "Age of the Terrible Lizards"

39. Time in the Fossil Record when dinosaur bones were laid down
    (a) Dawn of Life
    (b) Pre-Cambrian Era
    (c) Paleozoic Era
    (d) Cosmic Creation
    (e) Mesozoic Era

40. Possible explanation offered for the relatively sudden extinction of the dinosaurs
    (a) Law of Hidden Evil
    (b) Might makes right!
    (c) Impact hypothesis
    (d) Cosmic Consciousness
    (e) Alien invasion hypothesis

41. Appearance of Homo sapiens
    (a) About 400,000 to 1 million years ago
    (b) The Ice Age
    (c) Carboniferous Era
    (d) Stromatolite Period
    (e) Fish-like Age

42. Monkeys, apes, and humans
    (a) Amphibians
    (b) Reptiles
    (c) Urochordates
    (d) Primates
    (e) Pterosaurs

43. Stated the Theory of Evolution by Natural Selection
    (a) James Watson
    (b) Charles O'Connor
    (c) Charles Darwin
    (d) Gregor Mendel
    (e) Francis Crick

44. By adaptation, it is meant that
    (a) The external environment provides a larger copy which animal genes follow
    (b) Certain patterns of Biological Order make particular organisms "more fit" to survive in a given environment
    (c) A gradual process of destruction is unfolding
    (d) Natural Selection causes all body structures to deteriorate
    (e) "Eat-or-be eaten" always prevails!

45. Main builders of order or pattern at the chemical level
    (a) Organelles
    (b) Glucose molecules
    (c) Chemical bonds
    (d) Amino groups
    (e) Cell autolyzers

46. "Like dissolves like" suggests that
    (a) Solutions can be composed of any combination of contributing particles
    (b) Solvents dissolve solutes with similar bonding and/or electrical charge to their own
    (c) Solvents will just break down solutions with oppositely charged particles
    (d) Solutes, not solvents, really "like" to do the "dissolving"!
    (e) Neither solutes nor solvents can combine to produce solutions

47. A substance that breaks down into ions when placed into water
    (a) Carbohydrate
    (b) Electrolyte
    (c) Any compound
    (d) Organic molecule
    (e) Hydrocarbon

48. Carbon atoms create highly orderly structural skeletons because they
    (a) Easily mix with both NaCl and H_2O
    (b) Readily form covalent bonds with one another
    (c) Push water molecules against one another to force them to create mist
    (d) Exist in the liquid state
    (e) Usually diffuse away without leaving any part of them behind

49. Proteins that speed up many chemical reactions
    (a) Antibodies
    (b) Hemoglobins
    (c) Enzymes
    (d) Lipids
    (e) C–C complexes

50. A group of fats and fat-like hydrocarbons that are not soluble in water
    (a) Insulins
    (b) Carbohydrates
    (c) Vitamins

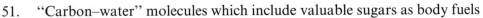

(d) Lipids
(e) Proteins

51. "Carbon–water" molecules which include valuable sugars as body fuels
    (a) Nucleic acids
    (b) Actin filaments
    (c) Lipids
    (d) ATPs
    (e) Carbohydrates

52. _____ is an important type of simple sugar molecule that is found in the human bloodstream
    (a) Glucose
    (b) Glycogen
    (c) Glucagon
    (d) Glucosamine
    (e) Pancreatic hydrase

53. The _____ acids derive their name from their occurrence within the cell "kernel"
    (a) Fatty
    (b) Amino
    (c) Nucleic
    (d) Nitrogenous
    (e) Acetic

54. An abbreviation for ribonucleic acid
    (a) Ribonuca
    (b) RNA
    (c) RBNA
    (d) RBA
    (e) Rib-Nucle-Ac

55. Occurs as a twisted ladder or double helix, with genes occurring along its length
    (a) RNA
    (b) Deoxyribose
    (c) Deoxyribonucleic acid
    (d) Viral protein coats
    (e) A film of fungi

56. The ability to do work
    (a) Metabolism
    (b) Differentiation

(c) Diffusion
(d) Mitosis
(e) Energy

57. Free energy that helps particles "move"
(a) Potential
(b) Stored
(c) ATP
(d) Enzymatic
(e) Kinetic

58. High-energy phosphate bonds within the ATP molecule
(a) Non-reducing
(b) Potential energy storage
(c) ATPases
(d) Nucleic
(e) Aerobic

59. A continual process between energy storage and energy release that occurs in cells
(a) Carbon–glucose conversion
(b) Hydrogen and amino acid transfer
(c) ATP–ADP Cycle
(d) Cellular phagocytosis
(e) Active immunity

60. A "condition of building up"
(a) Anabolism
(b) Glycolysis
(c) Catabolism
(d) Metabolism
(e) Egestion

61. Heterotroph
(a) An organism that produces its own energy
(b) Creature that consumes other organic foodstuffs for its nourishment
(c) Multicellular, free-living plant
(d) A type of molecule that is different from others
(e) An organism that feeds upon its own flesh!

62. The "process of breaking down sweets"
(a) Glycolysis
(b) Calvin Cycle

(c) Hydrolysis
(d) Photosynthesis
(e) Dehydration

63. Cellular respiration is
    (a) Usually anaerobic
    (b) Sometimes fatal
    (c) Aerobic breakdown of organic molecules
    (d) Always the same as photosynthesis
    (e) Totally dependent upon chlorophyll

64. Krebs Cycle
    (a) Occurs along the cristae of mitochondria
    (b) Takes place in the animal cell nucleus
    (c) Breaks down the pyruvic acids produced by glycolysis
    (d) Is named for Kathy Krebs
    (e) Results in 55 ATPs per glucose molecule consumed

65. "The cell is the basic unit of all living things"
    (a) Phlogiston Theory
    (b) Cloning Hypothesis
    (c) Krebs Ideology
    (d) Mendel's Model
    (e) Modern Cell Theory

66. "Colored bodies" containing DNA
    (a) Genes
    (b) Nuclei
    (c) Mitochondria
    (d) Chromosomes
    (e) Lysosomes

67. During the process of transcription
    (a) DNA codons are sequentially destroyed
    (b) Glucose is converted into CO_2 and water
    (c) Exposed DNA codons are used to make a messenger RNA molecule
    (d) Cell division is replaced by protein synthesis
    (e) Messenger RNA is matched with transfer RNA

68. Simple diffusion, osmosis, and facilitated diffusion are all classified as types of
    (a) Active transport
    (b) Pinocytosis

  (c) Respiratory systems
  (d) Passive transport systems
  (e) Phagocytosis

69. Phase of the Cell Cycle that occurs "between" cell divisions
  (a) Meiosis
  (b) Interphase
  (c) Prophase
  (d) Metaphase
  (e) Telophase

70. The division of paired, duplicated chromosomes into single, identical, unpaired chromosomes
  (a) Cephalization
  (b) Mitosis
  (c) Melanomiasis
  (d) Meiosis
  (e) Facilitated diffusion

71. All single-celled organisms belong to either of these two kingdoms
  (a) Animalia or Plantae
  (b) Fungi or Plantae
  (c) Protista or Monera
  (d) Bacteria or Plantae
  (e) Monera or Animalia

72. Usually consists of two or more species belonging to the same "stock or kind"
  (a) Population
  (b) Genus
  (c) Family
  (d) Order
  (e) Class

73. A kingdom consists of a group of related
  (a) Cousins
  (b) Organelles
  (c) Plants
  (d) Relatives
  (e) Phyla

74. A virus
  (a) Living sphere-or-rod-shaped organism
  (b) Part of every living cell

(c) Important component of the mitochondrion

(d) Helps its host cell survive and gain energy

(e) Non-living superchemical that parasitizes and thus "poisons" cells

75. In a condition of endosymbiosis, two organisms
    (a) Of different species live together, one inside of the body of the other
    (b) Survive and reproduce completely separately
    (c) Fight each other to exist!
    (d) Are completely passive and totally unaffected by the other's existence
    (e) Bond together as sexual pair mates

76. The fruiting body of a fungus
    (a) Mushroom
    (b) Spores
    (c) Hyphae
    (d) Basidia
    (e) Gonada

77. The vascular plants or tracheophytes
    (a) Have bodies containing hollow tubes and dense patterns of leaf veins
    (b) Bryophytes
    (c) Spongy cap mushrooms
    (d) Monera
    (e) Lack internal vessels or leaf veins

78. Gymnosperms
    (a) Mosses and liverworts
    (b) Sperms that work out in a gym!
    (c) Liverleaf or hepatica
    (d) Ferns with fronds
    (e) Vascular plants with "naked seeds"

79. _____ means "a conditioning of lessening," wherein one parent cell divides into two daughter cells, each possessing only half its number of chromosomes
    (a) Meiosis
    (b) Microspore packing
    (c) Pollination
    (d) Mitosis
    (e) Halitosis

80. The "animals" whose bodies contain "true" tissues
    (a) Parazoans
    (b) Protozoans
    (c) Corollas
    (d) Eumetazoans
    (e) Mitochondria

81. The lobster body is said to have a high degree of bilateral symmetry, because
    (a) It lacks a stiffening backbone
    (b) Two wrongs don't make a right!
    (c) It has a rough balance of various "rays" that project from the same center
    (d) The right and left halves are essentially mirror images of one another
    (e) Both sides never seem to "measure together" in the same way

82. The "outer skin" covering the gastrula
    (a) Endoderm
    (b) Cristae
    (c) Epidermis
    (d) Ectoderm
    (e) Mesoderm

83. The coelomates (such as earthworms)
    (a) Possess bodies with no central cavity
    (b) Always breathe with two lungs
    (c) Have bodies containing a central cavity around the digestive tract
    (d) Usually have a visceral mass plus mantle
    (e) Include echinoderms with soft, smooth skin

84. Invertebrates with jointed bodies and "jointed feet"
    (a) Arthropods
    (b) Sea slugs
    (c) Clams and oysters
    (d) Petunia plants
    (e) Common plantain

85. Spiders and their relatives
    (a) Bivalves
    (b) Arachnids
    (c) Crustaceans
    (d) Chilopods
    (e) Diplopods

86. There are more species of _____ than species of all other types of animals, combined
    (a) Trout
    (b) Mollusks
    (c) Trilobites
    (d) Insects
    (e) Ants

87. An ancient forerunner of the vertebral column that supports and stiffens the body
    (a) Cranium
    (b) Notochord
    (c) Umbilical cord
    (d) Spine
    (e) Urochord

88. Literally, this Class is for the "birds"
    (a) Aves
    (b) Reptilia
    (c) Agnatha
    (d) Placodermi
    (e) Amphibia

89. Outermost layer of the skin or integument
    (a) Dermis
    (b) Subdermis
    (c) Osteoid
    (d) Epidermis
    (e) Keratin

90. Thermoregulation
    (a) Technical term for widening of vessels
    (b) Disruption of homeostasis
    (c) Another name for radiation
    (d) A process of bone support
    (e) Control of body heat or temperature

91. In humans, the _____ comprises many bones and joints
    (a) Endoskeleton
    (b) Carapace
    (c) Exoskeleton
    (d) Fibrous connective tissue
    (e) Endomysium

92. Ossification
    (a) Muscular contraction
    (b) Process of bone formation
    (c) Elimination of unusable food residue
    (d) Respiration plus ventilation
    (e) Snapping of intervertebral joints

93. Current explanation of how muscles produce body movement
    (a) Law of Diminishing Returns
    (b) Sliding Filament Theory
    (c) Neuromuscular Junction
    (d) Tales of the epimysium
    (e) Motor Pathway Formation

94. The hypothalamus of the brain contains a number of _____ for homeostasis
    (a) Visceral effectors
    (b) Proprioceptors
    (c) Motor neurons
    (d) Control centers
    (e) Primary visual areas

95. Glands of "internal" secretion of hormones into the bloodstream
    (a) Exocrine
    (b) Target cells
    (c) Endocrine
    (d) Sweat
    (e) Sympathetic

96. Organ system that includes the heart, blood, and blood vessels
    (a) Pulmonary
    (b) Digestive
    (c) Genitourinary
    (d) Cardiovascular
    (e) Lymphatic-Immune

97. The reticuloendothelial (R-E) system
    (a) The lymph nodes plus thymus gland
    (b) A "little network" of lymphatic vessels lined by flat, scale-like cells, along with a collection of related organs
    (c) Wandering macrophages and red bone marrow
    (d) Just a bunch of dead-ended lymphatic capillaries
    (e) The spleen and its attachments

98. Pulmonary ventilation differs from respiration in that
    (a) Pulmonary ventilation is the sum of inspiration plus expiration, whereas respiration is simply gas exchange between various body compartments
    (b) Pulmonary ventilation involves both gas exchange and removal of stale air from the body
    (c) Respiration only involves gas exchange between the air in the lung alveoli and the blood in the pulmonary capillaries
    (d) Pulmonary ventilation only involves the upper respiratory pathway lying above both lungs
    (e) Inspiration, but not ventilation, requires the bulk flow of air

99. Digestion differs from absorption in the following way
    (a) Digestion is the movement of material from the interior of the digestive tube into the bloodstream, whereas absorption is the breakdown of foodstuffs
    (b) Digestion is the chemical and physical breakdown of food, whereas absorption is the movement of nutrient particles from the tube into the bloodstream
    (c) Absorption is the active adding of material to the digestive tube, from the bloodstream
    (d) Digestion is the same as secretion from the accessory digestive organs
    (e) Absorption is identical to both ingestion and egestion, whereas digestion is not identical to these processes

100. The genitourinary system
    (a) The reproductive organs and pathways in both males and females
    (b) Involves just the organs of urine formation and excretion
    (c) A collection of specialized cells that directly participate in body defense
    (d) Is located just inferior and dorsal to most of the neuromuscular system
    (e) Concept that the reproductive and urinary pathways share many traits

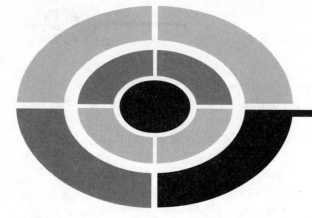

APPENDIX: Answers to Quiz, Test, and Final Exam Questions

Chapter 1

1. D	2. B	3. A	4. B
5. B	6. D	7. C	8. B
9. A	10. B		

Chapter 2

1. C	2. B	3. C	4. A
5. A	6. B	7. C	8. B
9. C	10. D		

Chapter 3

1. B	2. C	3. B	4. A
5. C	6. A	7. C	8. C
9. C	10. B		

APPENDIX: Answers

Test: Part 1

1. D	2. A	3. B	4. B
5. B	6. C	7. D	8. C
9. C	10. B	11. C	12. A
13. B	14. E	15. E	16. B
17. C	18. D	19. B	20. C
21. C	22. C	23. C	24. B
25. E			

Chapter 4

1. C	2. D	3. A	4. A
5. A	6. D	7. B	8. A
9. A	10. C		

Chapter 5

1. D	2. A	3. D	4. A
5. B	6. C	7. A	8. B
9. A	10. B		

Test: Part 2

1. B	2. C	3. D	4. A
5. B	6. C	7. A	8. E
9. A	10. B	11. E	12. B
13. A	14. A	15. A	16. B
17. D	18. A	19. E	20. D
21. A	22. B	23. A	24. D
25. A			

Chapter 6

1. B	2. A	3. C	4. C
5. A	6. B	7. A	8. D
9. D	10. B		

Chapter 7

1. C	2. B	3. A	4. B
5. A	6. A	7. B	8. C
9. C	10. B		

Chapter 8

1. C	2. C	3. B	4. C
5. D	6. B	7. C	8. A
9. B	10. B		

Chapter 9

1. A	2. D	3. B	4. A
5. C	6. B	7. D	8. A
9. A	10. C		

Chapter 10

1. C	2. C	3. A	4. D
5. C	6. D	7. B	8. A
9. D	10. C		

Chapter 11

1. B	2. D	3. B	4. C
5. D	6. A	7. C	8. D
9. A	10. C		

Chapter 12

1. D	2. B	3. B	4. B
5. D	6. B	7. A	8. C
9. B	10. C		

Test: Part 3

1. C	2. B	3. C	4. D
5. C	6. C	7. A	8. B
9. B	10. C	11. D	12. A
13. A	14. C	15. A	16. B
17. A	18. E	19. E	20. A
21. C	22. C	23. E	24. C
25. E			

Chapter 13

1. A	2. B	3. A	4. A
5. D	6. A	7. C	8. A
9. A	10. A		

Chapter 14

1. B	2. C	3. A	4. D
5. B	6. C	7. A	8. D
9. B	10. B		

APPENDIX: Answers

Chapter 15

1. C	2. D	3. D	4. C
5. D	6. B	7. D	8. D
9. C	10. C		

Chapter 16

1. D	2. D	3. A	4. B
5. C	6. C	7. C	8. A
9. A	10. D		

Chapter 17

1. B	2. B	3. D	4. D
5. B	6. A	7. D	8. C
9. A	10. D		

Chapter 18

1. D	2. C	3. D	4. D
5. C	6. C	7. C	8. D
9. D	10. C		

Chapter 19

1. C	2. A	3. A	4. D
5. C	6. C	7. B	8. C
9. C	10. D		

Chapter 20

1. B	2. C	3. A	4. D
5. C	6. D	7. C	8. D
9. C	10. A		

Test: Part 4

1. A	2. A	3. E	4. C
5. B	6. A	7. A	8. D
9. B	10. C	11. A	12. B
13. C	14. C	15. A	16. A
17. A	18. C	19. C	20. B
21. C	22. A	23. D	24. A
25. D			

Final Exam

1. B	2. A	3. D	4. E
5. A	6. C	7. A	8. B
9. D	10. E	11. C	12. D
13. C	14. B	15. A	16. E
17. B	18. E	19. C	20. C
21. B	22. D	23. A	24. C
25. A	26. C	27. B	28. A
29. C	30. D	31. C	32. C
33. E	34. C	35. D	36. B
37. E	38. B	39. E	40. C
41. B	42. D	43. C	44. B
45. C	46. B	47. B	48. B
49. C	50. D	51. E	52. A
53. C	54. B	55. C	56. E
57. E	58. B	59. C	60. A
61. B	62. A	63. C	64. C
65. E	66. D	67. C	68. D
69. B	70. B	71. C	72. B
73. E	74. E	75. A	76. A
77. A	78. E	79. A	80. D
81. D	82. D	83. C	84. A
85. B	86. D	87. B	88. C
89. D	90. E	91. A	92. B
93. B	94. D	95. C	96. D
97. B	98. A	99. B	100. E

INDEX

ABOUT THE AUTHOR

Dr. Dale Layman is a Professor of Biology, and Human Anatomy and Physiology, at Joliet Junior College. A resident of Joliet, Illinois, Dr. Layman is a frequent author with many international honors and awards. He has more than 28 years of experience in the field of biological sciences.